Holmes McDougall Metric Maths

BOOK 2

J. O'Neill
J. Potts
J. Wood

General Editor : I. D. Watt

Holmes McDougall Ltd

Art-work by F. Vaughan
Printed by Holmes McDougall Limited
© 1970 Holmes McDougall Reprinted and revised March 1972

Contents Book 2

Vina Kirkpatrick.

INTRODUCTORY NOTE

The whole atmosphere of education has changed in the post-war years, particularly in the last decade. Everywhere, at all stages, there is evident a quickening of interest and a lively urge to try out new ideas and new methods, to provide an education that makes sense to the children of to-day. Teachers are aware, as no-one else, that there is no room for complacency, that we are nearer the beginning than the end of educational development. What can be claimed is that the prospect of sensible and sustained advance is brighter than ever before.

Nowhere is change more evident than in the Primary School. The day of the stereotype in syllabuses and textbooks is quickly going. New subjects have been introduced, old ones refurbished and modernised. Teaching methods are placing greater emphasis on assignments and pupil participation. The stress is on adaptability and the capacity to think creatively. The demand for change has come from the schools themselves. The practising teacher in the classroom is becoming more and more the pacemaker and the initiator of new ideas of what to teach and how to teach.

Of all subjects in the Primary School curriculum, mathematics is the one that is undergoing the greatest change. Up until a few years ago the teaching of the subject was in large part confined to the development of skill in reckoning and the manipulation of unwieldy computations of little relevance and doubtful educational value. Much of the work was of a routine nature, much of it divorced from practical realities, with little understanding of the nature of the underlying concepts. This is being changed.

Materials for the study of mathematics—number, quantity and shape—abound in the child's environment and, by means of meaningful experiences, purposefully planned, children are being led progressively to the understanding of basic mathematical concepts. But, as the Memorandum on Primary Education in Scotland points out, there is no suggestion that all the material to be studied can be found directly in the environment or that none of the subjects grouped under Environmental Studies has to be pursued as a separate discipline. In the words of the Memorandum: "Indeed, in the case of mathematics, especially that part of it which is concerned with number, there must be from the earliest stages training in specific skills that will enable the pupils to handle quickly and efficiently the mathematical situations which will arise in the course of their other activities."

In this series of development exercises it is the need for this basic training in specific skills that the authors—all experienced in the day-to-day work of the schools—have had constantly in mind. Drafts of the exercises were tried out in a variety of schools and modified in the light of the reactions of both teachers and pupils. The result is a series of books which, it is confidently expected, will make a distinctive and useful contribution to the teaching of mathematics in primary schools.

Do You Know Book 1?

$26 + 48 + 25$

$26p + 48p + 25p$

Add up: (Check your answers by adding down)

1

(1) 11 72	(2) 33 31	(3) 82 23	(4) 21p 61p	(5) 41cm 14cm	(6) 2p 44p
(7) 15 31	(8) 41 68	(9) 72 25	(10) 52p 42p	(11) 86cm 4cm	(12) 41p 37p
(13) 19 7 87	(14) 92 26 35	(15) 277 97 326	(16) 18p 34p 24p	(17) 7cm 45cm 25cm	(18) 16p 28p 47p
(19) 190 248 458	(20) 526 239 187	(21) 474 213 75	(22) 19p 9p 38p	(23) 18cm 36cm 26cm	(24) 23p 19p 46p

$64cm − 28cm$

$64 − 28$

Subtraction: (Check answers by addition)

2

(1) 89 −31	(2) 57 −42	(3) 74 −52	(4) 65p −23p	(5) 98cm −84cm	(6) 42p −31p
(7) 26 −25	(8) 110 −70	(9) 90 −72	(10) 33p −10p	(11) 91cm −60cm	(12) 88p −62p
(13) 150 −83		(14) 110 −7	(15) 90p −64p		(16) 120cm −96cm
(17) 252 −114		(18) 473 −367	(19) 160p −71p		(20) 81cm −46cm

DO YOU KNOW BOOK 1?

The Multiplication tables:

X	0	1	2	3	4	5	6	7	8	9	10
1	0	1	2	3	4	5	6	7	8	9	10
2	0	2	4	6	8	10	12	14	16	18	20
3	0	3	6	9	12	15	18	21	24	27	30
4	0	4	8	12	16	20	24	28	32	36	40
5	0	5	10	15	20	25	30	35	40	45	50
6	0	6	12	18	24	30	36	42	48	54	60
10	0	10	20	30	40	50	60	70	80	90	100

First Factor

The product of 6 and 4 is 24
4 and 6 are called *factors* of 24
6 is the *quotient* on dividing 24 by 4
4×6 when read as '4 times 6' means

⊡ ⊡ ⊡ ⊡ or 4 groups of 6
4×6 when read as '4 multiplied by 6' means

⊡ ⊡ ⊡ ⊡ ⊡ ⊡ or 6 groups of 4.

3

Find the products:

(1) 2×3 (2) 2×6 (3) 3×5 (4) 4×4 (5) 5×3
(6) 6×2 (7) 3×7 (8) 2×9 (9) 10×4 (10) 6×6
(11) 2×1 (12) 4×6 (13) 5×8 (14) 6×0 (15) 3×8
(16) 4×9 (17) 6×8 (18) 10×7 (19) 5×9 (20) 10×10

4

Find the quotients:

(1) $10 \div 5$ (2) $18 \div 6$ (3) $12 \div 3$ (4) $25 \div 5$ (5) $14 \div 2$
(6) $32 \div 4$ (7) $27 \div 3$ (8) $8 \div 2$ (9) $90 \div 10$ (10) $0 \div 4$
(11) $30 \div 6$ (12) $42 \div 6$ (13) $80 \div 10$ (14) $50 \div 5$ (15) $28 \div 4$
(16) $54 \div 6$ (17) $30 \div 5$ (18) $18 \div 3$ (19) $16 \div 2$ (20) $60 \div 6$

5

Find the answers to:

(1) $(2 \times 4) + 3$ (2) $(3 \times 3) + 1$ (3) $(4 \times 5) - 2$
(4) $(5 \times 6) + 4$ (5) $(6 \times 1) - 4$ (6) $(10 \times 2) - 1$
(7) $(3 \times 9) + 6$ (8) $(2 \times 5) - 3$ (9) $(10 \times 8) + 2$
(10) $(5 \times 7) + 6$ (11) $(6 \times 5) - 3$ (12) $(4 \times 8) - 7$
(13) $5 + (4 \times 7)$ (14) $7 + (3 \times 6)$ (15) $2 + (6 \times 3)$
(16) $3 + (5 \times 4)$ (17) $10 - (6 \times 0)$ (18) $20 - (3 \times 4)$
(19) $6 + (10 \times 5)$ (20) $8 - (5 \times 1)$

6

Division:

(1) $2\,\overline{)\,48}$ (2) $2\,\overline{)\,80}$ (3) $2\,\overline{)\,92}$ (4) $2\,\overline{)\,77}$
(5) $3\,\overline{)\,69}$ (6) $3\,\overline{)\,93}$ (7) $3\,\overline{)\,75}$ (8) $3\,\overline{)\,86}$
(9) $4\,\overline{)\,88}$ (10) $4\,\overline{)\,92}$ (11) $4\,\overline{)\,76}$ (12) $4\,\overline{)\,63}$
(13) $5\,\overline{)\,55}$ (14) $5\,\overline{)\,65}$ (15) $5\,\overline{)\,90}$ (16) $5\,\overline{)\,82}$
(17) $6\,\overline{)\,66}$ (18) $6\,\overline{)\,84}$ (19) $6\,\overline{)\,90}$ (20) $6\,\overline{)\,99}$

Number Sentences

Copy and write the correct sign (+ or −) for each △:

7

(1) $8 = 6 \triangle 2$ (2) $2 \triangle 6 = 8$
(3) $8 \triangle 6 = 2$ (4) $15 = 7 \triangle 8$
(5) $15 \triangle 8 = 7$ (6) $8 \triangle 7 = 15$
(7) $20 = 10 \triangle 10$ (8) $10 \triangle 10 = 0$
(9) $10 + 4 = 9 \triangle 5$ (10) $10 + 3 = 20 \triangle 7$
(11) $8 \triangle 9 = 10 + 7$ (12) $19 \triangle 3 = 8 + 8$
(13) $4 + (3 + 7) = 20 \triangle 6$ (14) $6 \triangle 5 = 10 \triangle 1$
(15) $32 + 8 = 50 \triangle 10$ (16) $9 \triangle 6 = (16 - 6) + 5$
(17) $79 = 56 \triangle 23$ (18) $79 \triangle 56 = 23$
(19) $79 \triangle 23 = 56$ (20) $79 = 23 \triangle 56$

$13 - 8 = 5$ is a subtraction sentence

$5 + 8 = 13$ is an addition sentence

From add to subtract

$5 + 8 = 13$
↓
$13 - 5 = 8$
and $13 - 8 = 5$

Write each addition sentence as 2 subtraction sentences:

8

(1) $4 + 5 = 9$ (2) $6 + 2 = 8$ (3) $3 + 7 = 10$
(4) $2 + 13 = 15$ (5) $8 + 6 = 14$ (6) $7 + 11 = 18$
(7) $9 + 8 = 17$ (8) $6 + 10 = 16$ (9) $5 + 18 = 23$
(10) $6 + 21 = 27$ (11) $18 + 4 = 22$ (12) $21 + 9 = 30$
(13) $15 + 12 = 27$ (14) $14 + 11 = 25$ (15) $21 + 30 = 51$
(16) $63 + 9 = 72$ (17) $24 + 34 = 58$ (18) $29 + 50 = 79$
(19) $41 + 35 = 76$ (20) $53 + 44 = 97$

From subtract to add

$15 - 6 = 9$
↓
$6 + 9 = 15$
$9 + 6 = 15$

Write each subtraction sentence as 2 addition sentences:

9

(1) $7 - 5 = 2$ (2) $6 - 2 = 4$ (3) $8 - 1 = 7$
(4) $9 - 3 = 6$ (5) $12 - 4 = 8$ (6) $11 - 6 = 5$
(7) $14 - 4 = 10$ (8) $18 - 5 = 13$ (9) $20 - 9 = 11$
(10) $26 - 12 = 14$ (11) $30 - 10 = 20$ (12) $49 - 26 = 23$
(13) $52 - 30 = 22$ (14) $64 - 14 = 50$ (15) $48 - 25 = 23$
(16) $68 - 16 = 52$ (17) $77 - 35 = 42$ (18) $80 - 11 = 69$
(19) $40 - 14 = 26$ (20) $87 - 19 = 68$

Words to Symbols

Words: Seventeen plus five equals twenty-two

Symbols: 17   +   5   =   22

Words: Take six from fifteen and the result is nine

Symbols: $15 - 6 = 9$

Write these sentences using symbols:

10

(1) Six plus seven equals thirteen.

(2) Twelve less five equals seven.

(3) Add 9 and 8 and the answer is 17.

(4) On adding 9 and 14 the answer is twenty-three.

(5) Take 12 from 20 and the result is 8.

(6) From 10 take away 1 and the answer is nine.

(7) Twenty plus thirty equals fifty.

(8) Forty taken from ninety gives fifty.

(9) Twenty two less six equals sixteen.

(10) Ten added to fourteen equals twenty-four.

Find $\square$:

11

(1) $8 + 8 = \square$	(2) $11 - 2 = \square$	(3) $4 + 6 = \square$
(4) $23 - 6 = \square$	(5) $\square - 3 = 8$	(6) $\square - 9 = 6$
(7) $9 + \square = 18$	(8) $\square + 1 = 13$	(9) $13 - \square = 8$
(10) $7 + \square = 9$	(11) $\square + 15 = 18$	(12) $\square - 10 = 35$
(13) $14 + \square = 21$	(14) $32 - \square = 11$	(15) $\square - 7 = 19$
(16) $4 + \square = 19$	(17) $14 - \square = 7$	(18) $22 + \square = 47$
(19) $\square + 41 = 84$	(20) $64 - \square = 14$	

Using Brackets

Abe has found an easier way to add some numbers.

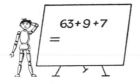

$$13 + (18+2)$$
$$= 13+20$$
$$= 33$$

The numbers in brackets add up to tens

$13 + 18 + 2$

Copy each example.

Use brackets like Abe to find the answers:

12

(1) $(3 + 7) + 9$	(2) $6 + (8 + 2)$	(3) $(9 + 11) + 7$
(4) $2 + 18 + 7$	(5) $8 + 15 + 5$	(6) $18 + 2 + 6$
(7) $9 + 17 + 3$	(8) $19 + 1 + 18$	(9) $24 + 6 + 20$
(10) $10 + 8 + 72$	(11) $16 + 53 + 7$	(12) $71 + 9 + 5$
(13) $90 + 10 + 80$	(14) $91 + 9 + 83$	(15) $99 + 93 + 7$

$63 + 9 + 7$
$=$

It appears that Abe can not do this in the same way.

$63 + 9 + 7$

LOOK!

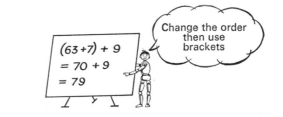

$$(63 + 7) + 9$$
$$= 70 + 9$$
$$= 79$$

Change the order then use brackets

Change the order of each.

Use the brackets like Abe to find the answers:

13

(1) $8 + 9 + 2$	(2) $6 + 11 + 4$	(3) $1 + 14 + 9$
(4) $7 + 15 + 3$	(5) $11 + 14 + 9$	(6) $5 + 16 + 15$
(7) $12 + 19 + 8$	(8) $18 + 21 + 2$	(9) $16 + 27 + 4$
(10) $5 + 24 + 25$	(11) $3 + 26 + 17$	(12) $8 + 23 + 12$
(13) $14 + 12 + 6$	(14) $95 + 21 + 5$	(15) $7 + 56 + 93$
(16) $5 + 12 + 5$	(17) $13 + 8 + 7$	(18) $19 + 22 + 1$
(19) $3 + 18 + 17$	(20) $8 + 10 + 72$	

Magic Squares

2	7	6
9	5	1
4	3	8

14

(1) Copy this square with numbers into your jotter.

(2) What is the sum of the numbers 2, 7 and 6 in the top row?

(3) Add the numbers in the second row.

(4) Add the numbers in the bottom row.

(5) What is the sum of the numbers 2, 9, and 4 in the left-hand column?

(6) Add the numbers 7, 5 and 3 in the second column.

(7) Add the numbers 6, 1 and 8 in the right-hand column.

(8) Add the numbers 2, 5 and 8 which lie on a diagonal.

(9) Add the numbers 4, 5 and 6 in the other diagonal.

(10) Abe thinks this is magic. Why?

In a magic square, the numbers are such that each line adds up to the same number.

Copy and complete the magic squares:

15

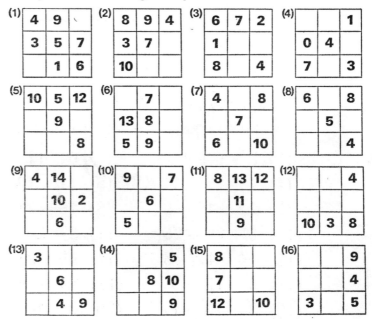

(1)
4	9	
3	5	7
	1	6

(2)
8	9	4
3	7	
10		

(3)
6	7	2
1		
8		4

(4)
		1
0	4	
7		3

(5)
10	5	12
	9	
		8

(6)
	7	
13	8	
5	9	

(7)
4		8
	7	
6		10

(8)
6		8
	5	
		4

(9)
4	14	
	10	2
	6	

(10)
9		7
	6	
5		

(11)
8	13	12
	11	
	9	

(12)
		4
10	3	8

(13)
3		
	6	
	4	9

(14)
		5
	8	10
		9

(15)
8		
7		
12		10

(16)
		9
		4
3		5

Telling the time

60 Minutes = 1 Hour.

The large hand or minute hand travels right round the clock face in 1 hour.

To make it easier to tell the time most clock faces have minutes marked round the edge.

One way of "telling the time" is to say the hour and the number of minutes (as shown by the minute hand) past the hour:

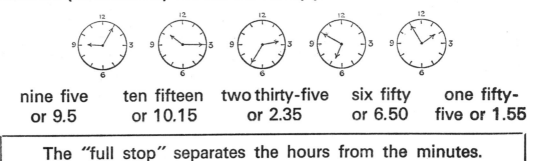

| nine five or 9.5 | ten fifteen or 10.15 | two thirty-five or 2.35 | six fifty or 6.50 | one fifty-five or 1.55 |

The "full stop" separates the hours from the minutes.

Note: There are five minutes between the hour markings.

Abe's clock shows the number of the minutes marked round the outside of the clock-face.

The time on Abe's clock is written:

9.37

This means 37 minutes past 9.

Write down the time on each of the clock faces below :

(1)   (2)   (3)   (4)   (5)

(6)   (7)   (8)   (9)   (10)

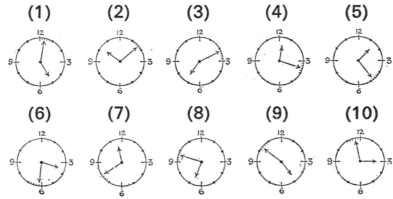

20 minutes past 3 = 3.20

17 Write the following times in this way:
(1) 25 past 6   (2) 5 past 4   (3) 10 past 8   (4) 20 past 12
(5) 27 minutes past 9   (6) a quarter past 5   (7) half past 10

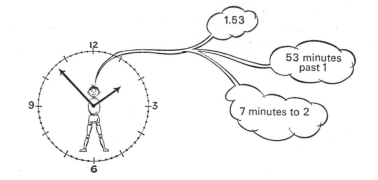

Note: If the number of minutes *past* is more than 30, the time may
be read as a number of minutes *to* the next hour.

For example, 5.40 = 40 minutes past 5 = 20 minutes to 6
minutes *past*   +   minutes *to* = 60
40   +   20   = 60

18 Write the following times using the words "minutes to":
(1) 2.40   (2) 7.40   (3) 12.40   (4) 40 minutes past 4
(5) 50 minutes past 6   (6) 3.50   (7) 10.50   (8) 1.45
(9) a quarter to 7   (10) 45 minutes past 9   (11) 9.55
(12) 5.55   (13) 55 minutes past 12   (14) 4.35   (15) 8.35
(16) 35 minutes past 11   (17) 11.59   (18) 54 minutes past 5
(19) 3.43   (20) 31 minutes past 5   (21) 9.46   (22) 2.34

10 minutes to 7 = 6.50

Write the following times in this way:

19 (1) 20 to 9   (2) 5 to 1   (3) 25 to 12   (4) 10 to 3
(5) a quarter to 7   (6) 17 minutes to 4   (7) 23 minutes to 11
(8) 19 minutes to 4   (9) 21 minutes to 5   (10) 2 minutes to 12
(11) 1 minute to 1   (12) 11 minutes to 10

Number Sentences

20 Copy and replace △ by the correct sign (× and ÷):

  (1) 2 △ 2 = 4   (2) 5 △ 4 = 20   (3) 9 △ 3 = 3
  (4) 20 △ 10 = 2   (5) 6 △ 4 = 24   (6) 35 △ 5 = 7
  (7) 4 △ 10 = 40   (8) 6 △ 6 = 1   (9) 25 △ 5 = 5
 (10) 3 △ 6 = 18   (11) 10 △ 5 = 50   (12) 12 △ 3 = 4
 (13) 6 △ 6 = 36   (14) 60 △ 6 = 10   (15) 20 △ 4 = 5
 (16) 10 △ 10 = 100   (17) 3 △ 3 = 1   (18) 3 △ 10 = 30
 (19) 20 △ 2 = 10   (20) 0 △ 4 = 0

21 Write each multiplication sentence as 2 division sentences:

 (1) $2 \times 3 = 6$ (2) $3 \times 5 = 15$
 (3) $4 \times 10 = 40$ (4) $2 \times 6 = 12$
 (5) $3 \times 6 = 18$ (6) $4 \times 5 = 20$
 (7) $6 \times 10 = 60$ (8) $5 \times 6 = 30$
 (9) $10 \times 2 = 20$ (10) $10 \times 5 = 50$

Write each division sentence as 2 multiplication sentences:

 (11) $8 \div 4 = 2$ (12) $12 \div 3 = 4$
 (13) $30 \div 10 = 3$ (14) $10 \div 5 = 2$
 (15) $24 \div 6 = 4$ (16) $45 \div 5 = 9$
 (17) $42 \div 6 = 7$ (18) $32 \div 4 = 8$
 (19) $54 \div 6 = 9$ (20) $40 \div 5 = 8$

22 Find ☐:

  (1) $3 \times 8 = \square$ (2) $16 \div 2 = \square$ (3) $5 \times 7 = \square$
  (4) $27 \div 3 = \square$ (5) $\square \div 4 = 8$ (6) $\square \div 5 = 5$
  (7) $2 \times \square = 14$ (8) $\square \times 9 = 18$ (9) $21 \div \square = 7$
 (10) $3 \times \square = 9$ (11) $\square \times 7 = 28$ (12) $\square \div 4 = 4$
 (13) $4 \times \square = 36$ (14) $45 \div \square = 9$ (15) $\square \div 6 = 7$
 (16) $5 \times \square = 40$ (17) $54 \div \square = 9$ (18) $6 \times \square = 36$
 (19) $\square \times 8 = 48$ (20) $100 \div \square = 10$

Words to Symbols

23 Write these sentences using symbols:

  (1) Three times nine equals twenty-seven.
  (2) Fourteen divided by two equals seven.
  (3) Thirty divided by five is equal to six.
  (4) Eight times five is equal to forty.
  (5) Six multiplied by six is equal to thirty-six.
  (6) Ninety divided by ten is equal to nine.
  (7) Four times twenty is equal to eighty.
  (8) Sixty divided by two equals thirty.
  (9) Forty multiplied by five equals two hundred.
(10) Thirty-six divided by three equals twelve.

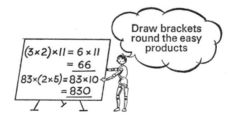

$3 \times 2 \times 11$
$83 \times 2 \times 5$

Copy each example.

24 Use brackets like Abe to find the answers:

(1) $(2 \times 3) \times 9$	(2) $(2 \times 2) \times 7$	(3) $8 \times (2 \times 2)$
(4) $2 \times 2 \times 9$	(5) $7 \times 2 \times 3$	(6) $2 \times 3 \times 8$
(7) $9 \times 2 \times 3$	(8) $3 \times 2 \times 7$	(9) $8 \times 3 \times 2$
(10) $2 \times 5 \times 9$	(11) $2 \times 5 \times 7$	(12) $8 \times 5 \times 2$
(13) $10 \times 5 \times 2$	(14) $2 \times 5 \times 14$	(15) $5 \times 2 \times 17$
(16) $21 \times 5 \times 2$	(17) $5 \times 2 \times 41$	(18) $38 \times 2 \times 5$
(19) $29 \times 2 \times 5$	(20) $5 \times 2 \times 32$	

Change the order of each. Use brackets like Abe to find the **answers:**

25
(1) $2 \times 9 \times 5$ (2) $5 \times 7 \times 2$ (3) $2 \times 8 \times 5$ (4) $5 \times 12 \times 2$
(5) $5 \times 15 \times 2$ (6) $2 \times 18 \times 5$ (7) $5 \times 22 \times 2$ (8) $2 \times 27 \times 5$
(9) $2 \times 36 \times 5$ (10) $5 \times 43 \times 2$ (11) $2 \times 48 \times 5$ (12) $5 \times 52 \times 2$
(13) $5 \times 61 \times 2$ (14) $2 \times 66 \times 5$ (15) $5 \times 69 \times 2$ (16) $5 \times 72 \times 2$
(17) $2 \times 80 \times 5$ (18) $2 \times 84 \times 5$ (19) $5 \times 89 \times 2$ (20) $2 \times 100 \times 5$

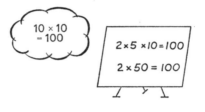

Write down the answers to:

26
(1) $2 \times 4 \times 10$ (2) 2×40 (3) $3 \times 3 \times 10$ (4) 3×30
(5) $4 \times 2 \times 10$ (6) 4×20 (7) $5 \times 3 \times 10$ (8) 5×30
(9) $6 \times 6 \times 10$ (10) 6×60 (11) $2 \times 8 \times 10$ (12) 2×80
(13) $3 \times 5 \times 10$ (14) 3×50 (15) $5 \times 7 \times 10$ (16) 5×70
(17) $4 \times 9 \times 10$ (18) 4×90 (19) $6 \times 9 \times 10$ (20) 6×90
(21) 2×70 (22) 4×70 (23) 5×50 (24) 3×40 (25) 6×40
(26) 4×50 (27) 2×90 (28) 3×60 (29) 6×50 (30) 5×80

Abe remembers this and writes:

2×70  70×2

2×70 has the same value as 70×2.

Find the products:

27
(1) 50×5 (2) 2×60 (3) 90×4 (4) 5×40 (5) 5×50
(6) 30×6 (7) 20×5 (8) 6×80 (9) 70×6 (10) 60×5

Copy and complete:

	(a)	(b)	(c)	(d)
28	(1) $3 \times 4 =$	$3 \times 5 =$	$3 \times (4 + 5) =$	$3 \times 9 =$
	(2) $2 \times 2 =$	$2 \times 8 =$	$2 \times (2 + 8) =$	$2 \times 10 =$
	(3) $4 \times 3 =$	$4 \times 7 =$	$4 \times (3 + 7) =$	$4 \times 10 =$
	(4) $6 \times 2 =$	$6 \times 6 =$	$6 \times (2 + 6) =$	$6 \times 8 =$
	(5) $5 \times 3 =$	$5 \times 4 =$	$5 \times (3 + 4) =$	$5 \times 7 =$
	(6) $4 \times 2 =$	$4 \times 10 =$	$4 \times (2 + 10) =$	$4 \times 12 =$
	(7) $2 \times 4 =$	$2 \times 10 =$	$2 \times (4 + 10) =$	$2 \times 14 =$
	(8) $5 \times 1 =$	$5 \times 20 =$	$5 \times (1 + 20) =$	$5 \times 21 =$
	(9) $3 \times 3 =$	$3 \times 30 =$	$3 \times (3 + 30) =$	$3 \times 33 =$
	(10) $6 \times 6 =$	$6 \times 40 =$	$6 \times (6 + 40) =$	$6 \times 46 =$

4×74 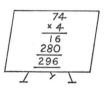 74×4

Multiplication:

28

(1) 81×2 (2) 74×2 (3) 22×3 (4) 73×3 (5) 62×4
(6) 5×91 (7) 6×61 (8) 3×63 (9) 6×51 (10) 6×91
(11) 42×4 (12) 24×2 (13) 41×6 (14) 2×51 (15) 37×2
(16) 79×3 (17) 3×66 (18) 4×78 (19) 5×33 (20) 6×57
(21) 6×12 (22) 25×3 (23) 96×5 (24) 55×4 (25) 56×2
(26) 42×4 (27) 43×3 (28) 5×79 (29) 34×4 (30) 68×6

Notation of Thousands

"Th" for Thousands

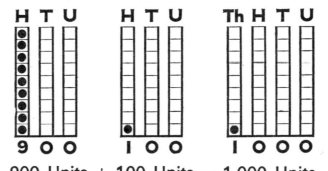

900 Units + 100 Units = 1 000 Units

1000 is read as ONE THOUSAND.

Abe's abacus shows 7
  in the thousands place.

This number is read as
  SEVEN THOUSAND.

This number is written 7 000

NOTE: The three zeros are grouped
  together and separated slightly
  from the seven.

30 (1) Read these numbers and write them in figures:
(a) six thousand   (b) one thousand   (c) eight thousand
(d) five thousand   (e) seven thousand (f) nine thousand

(2) Write the names of these numbers:
(a) 2 000 (b) 5 000 (c) 4 000 (d) 9 000 (e) 7 000 (f) 3 000

(3) Show each number on an abacus like Abe:
(a) four thousand   (b) 8 000   (c) five thousand   (d) 6 000
(e) three thousand (f) 1 000   (g) two thousand   (h) 9 000

READING AND WRITING NUMBERS

Th and U

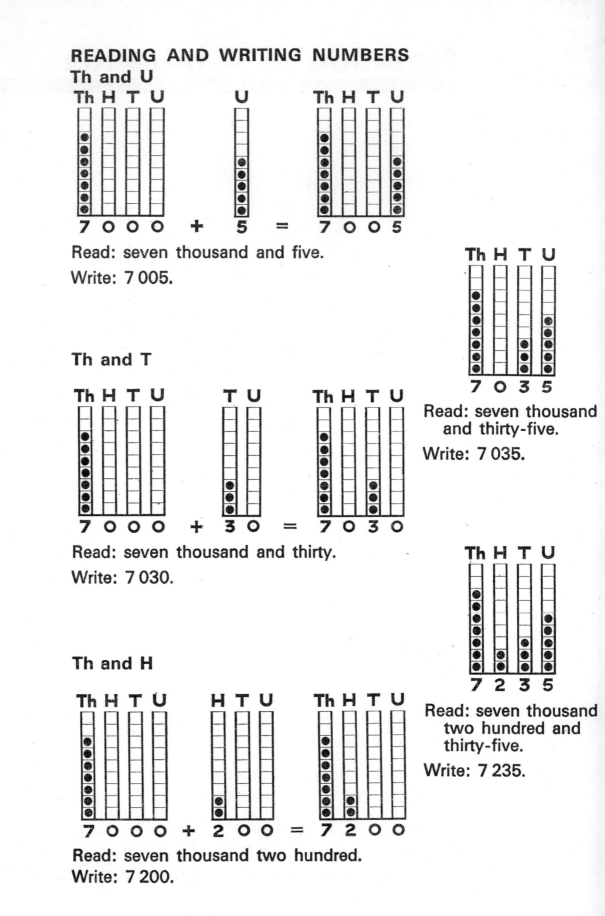

7 0 0 0   +   5   =   7 0 0 5

Read: seven thousand and five.
Write: 7 005.

7 0 3 5

Read: seven thousand
  and thirty-five.

Write: 7 035.

Th and T

7 0 0 0   +   3 0   =   7 0 3 0

Read: seven thousand and thirty.
Write: 7 030.

7 2 3 5

Read: seven thousand
  two hundred and
  thirty-five.

Write: 7 235.

Th and H

7 0 0 0   +   2 0 0   =   7 2 0 0

Read: seven thousand two hundred.
Write: 7 200.

(1) Write in figures:

31

(a) three thousand and seven   (b) one thousand and one
(c) three thousand and seventy (d) one thousand and ten
(e) three thousand and seventeen
(f) one thousand and seventy-one
(g) three thousand seven hundred
(h) seven thousand three hundred
(i) seven thousand one hundred and three
(j) one thousand three hundred and seven
(k) three thousand one hundred and seventeen
(l) one thousand seven hundred and seventy
(m) nine thousand three hundred and seventy-one

(2) Write the names of these numbers:

(a) 4 009   (b) 7 005   (c) 5 070   (d) 8 010   (e) 6 092
(f) 5 016   (g) 3 900   (h) 9 300   (i) 6 403   (j) 1 101
(k) 2 220   (l) 7 680   (m) 6 718   (n) 8 181   (o) 9 111

(3) Show each number on an abacus like Abe:

(a) 1 009   (b) two thousand and ninety
(c) three thousand and nineteen   (d) 4 900   (e) 5 109
(f) six thousand one hundred and ninety   (g) 7 919
(h) nine thousand and nine

(4) Write the answers:

(a) 999 + 1   (b) 8 999 + 1
(c) 990 + 10   (d) 8 990 + 10
(e) 900 + 100   (f) 8 900 + 100

(5) Using all four places on his abacus, Abe showed the largest number he could. This was,

"nine thousand nine hundred and ninety-nine."

Write this number in figures and show it on an abacus like Abe.

(6) Write in figures the number shown on each abacus:

(a)   (b)   (c)   (d)

Th H T U   Th H T U   Th H T U   Th H T U

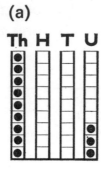 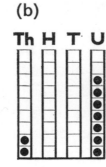 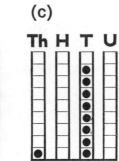

(e)   (f)   (g)   (h)

Th H T U   Th H T U   Th H T U   Th H T U

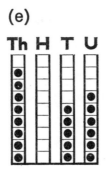

(i)   (j)   (k)   (l)

Th H T U   Th H T U   Th H T U   Th H T U

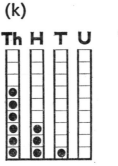 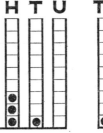

(m)   (n)   (o)   (p)

Th H T U   Th H T U   Th H T U   Th H T U

(7) Write the correct figure for each □ followed by its place name:

			thousands		hundreds		tens		units
(a)	5 024	means	5	"	0	"	2 tens and		4 units
(b)	4 052	"	4	"	0	"	□ "	"	2 "
(c)	2 045	"	□	"	0	"	4 "	"	□ "
(d)	7 012	"	7	"	□	"	□ "	"	2 "
(e)	3 010	"	3	"	0	"	□ "	"	□ "
(f)	6 007	"	□	"	0	"	□ "	"	7 "
(g)	9 019	"	9	"	□	"	□ "	"	9 "
(h)	9 099	"	9	"	□	"	□ "	"	□ "

(8) Write down the larger number from each pair:

(a) 9 019 and 8 099   (b) 9 019 and 9 099   (c) 7 012 and 7 021
(d) 6 048 and 5 084   (e) 1 023 and 2 013   (f) 8 027 and 2 078
(g) 6 078 and 8 076   (h) 1 009 and 1 090

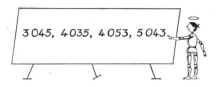

(9) Rearrange each list of numbers in order of size (smallest first):

(a) 1 023, 1 013, 1 033, 1 003   (b) 8 047, 5 047, 9 047, 2 047
(c) 9 018, 7 027, 5 036, 3 045   (d) 2 002, 1 090, 2 009, 1 019
(e) 9 090, 9 009, 9 019, 9 091   (f) 1 001, 1 010, 1 011, 1 000

Missing Figures

Copy down and fill in the missing figures in place of *:

(1)
```
  24
 +3*
 ───
  57
```

(2)
```
  *2
 +34
 ───
  96
```

(3)
```
  46
 +35
 ───
  *1
```

(4)
```
  26
 +33
 ───
  **
```

(5)
```
  52
 + *3
 ───
  8*
```

(6)
```
  **
 +42
 ───
  58
```

(7)
```
  *4
 +3*
 ───
  74
```

(8)
```
  1*
 +*4
 ───
  65
```

(9)
```
  **
 +38
 ───
  83
```

(10)
```
  65
 −42
 ───
  2*
```

(11)
```
  57
 −1*
 ───
  43
```

(12)
```
  86
 −*5
 ───
  5*
```

(13)
```
  78
 −**
 ───
  31
```

(14)
```
  **
 −35
 ───
  24
```

(15)
```
  36
 −*6
 ───
  2*
```

(16)
```
  84
 −16
 ───
  **
```

(17)
```
  90
 −  *
 ───
  85
```

(18)
```
  50
 −2*
 ───
  *6
```

(19)
```
  1*
 +*8
 ───
  37
```

(20)
```
  49
 +2*
 ───
  *5
```

(21)
```
  6
 +*7
 ───
  2*
```

(22)
```
  7*
 −*2
 ───
  30
```

(23)
```
  81
 −**
 ───
  40
```

(24)
```
  *6
 −1*
 ───
  47
```

Addition

ADDITION TO THOUSANDS

542 + 736 + 917

Abe gets 21 Hundreds.

Abe says, "Twenty-one. Write 1H in the answer and carry 2 thousands."

Note: 20 hundreds = 2 thousands.

Add:

33

(1)	332	(2)	951	(3)	872	(4)	263	(5)	246
	114		420		723		15		212
	543		25		602		810		737

(6)	662	(7)	103	(8)	832	(9)	902	(10)	81
	383		422		325		91		9
	900		254		403		83		837
			910		716		710		94

Find the answers to:

34

(1) 142 + 785 + 369   (2) 47 + 560 + 938   (3) 211 + 730 + 592

(4) 486 + 463 + 125   (5) 970 + 883 + 906   (6) 574 + 96 + 358

(7) 724 + 180 + 795   (8) 231 + 646 + 842   (9) 950 + 371 + 12

(10) 473 + 239 + 700 (11) 556 + 395 + 627 (12) 944 + 480 + 118

(13) 709 + 856 + 64 + 643   (14) 259 + 968 + 533 + 70

(15) 8 + 320 + 214 + 712   (16) 820 + 171 + 975 + 389

ADDITION WITH THOUSANDS $1\,032 + 2\,465 + 96 + 3\,127$

Abe adds and says,

"20 units. Write 0 and carry 2 tens
22 tens. Write 2 and carry 2 hundreds
7 hundreds. Write 7
6 thousands. Write 6"

35   Find the totals of these numbers:

(1) 1 325, 4 221 and 1 013   (2) 3 215, 4 262 and 1 423
(3) 3 413, 1 252 and 1 634   (4) 4 342, 823 and 2 414
(5) 317, 3 342 and 92   (6) 1 614, 465 and 3 841
(7) 2 357, 4 162 and 320   (8) 2 549, 4 362 and 2 151
(9) 1 215, 83 and 7 842   (10) 967, 1 331 and 2 563
(11) 2 158, 1 236 and 4 862   (12) 391, 2 214 and 283
(13) 3 368, 782 and 2 589   (14) 2 662, 3 853 and 763
(15) 1 773, 4 074 and 2 598   (16) 2 028, 1 706, 2 315 and 1 292
(17) 1 276, 2 188, 76 and 2 624   (18) 1 588, 2 149, 1 664 and 1 856
(19) 2 067, 2 658 and 3 982   (20) 1 247, 1 603, 2 367 and 1 099
(21) 2 597, 1 377, 1 248 and 1 063

36   Find the sums:

(1) 826 + 345 + 219   (2) 248 + 357 + 804
(3) 406 + 729 + 838   (4) 554 + 365 + 279
(5) 821 + 739 + 861   (6) 708 + 3 412 + 1 231
(7) 2 515+3 462+1 831+29   (8) 3 041+1 996+2 071+1 415
(9) 2 223+1 432+3 516+1 068   (10) 1 254+1 732+1 485+2 247
(11) Two hundred and ninety-four + Three thousand and seventy-six + Two thousand four hundred and eighty-seven.
(12) Two thousand six hundred and thirteen + One thousand five hundred and thirty-seven + One thousand eight hundred and five.

Subtraction

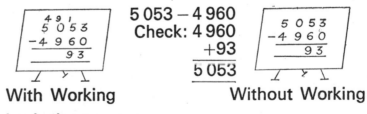

| Th H T U
 3̶4̶ ¹¹2̶ ¹⁷8̶ ¹3
 − 7 9 6
 3 4 8 7 | With
 Working | Without
 Working | 4·2·8·3
 − 7 9 6
 3 4 8 7 | 4 283 − 796 |

Subtract:

37

(1) 5 463 4 122	(2) 7 896 3 741	(3) 7 259 5 148	(4) 4 473 1 421
(5) 5 968 3 505	(6) 4 312 4 002	(7) 8 453 5 125	(8) 7 692 3 249
(9) 8 364 6 036	(10) 7 592 2 517	(11) 4 483 3 178	(12) 9 358 4 219
(13) 7 648 3 366	(14) 8 737 2 242	(15) 4 905 1 281	(16) 4 718 2 554

(17) 3 821 from 6 652 (18) 2 812 from 7 456 (19) 377 from 9 287
(20) 3 077 from 6 253 (21) 4 685 from 7 861 (22) 249 from 5 402
(23) 1 569 from 5 327 (24) 2 873 from 4 501 (25) 3 959 from 8 648
(26) 4 154 from 4 642 (27) 3 068 from 3 753 (28) 5 197 from 5 264

4 9 1
 5 0 5 3
 − 4 9 6 0
 9 3
	5 053 − 4 960
	 Check: 4 960
	    + 93
	   5 053
		5 0 5 3
		 − 4 9 6 0
		 9 3

With Working		**Without Working**

Find and check the answers:

38

(1) 3 697 − 2 413	(2) 5 483 − 3 172	(3) 6 561 − 6 430
(4) 4 232 − 1 117	(5) 5 967 − 2 328	(6) 4 831 − 4 213
(7) 5 451 − 316	(8) 1 085 − 39	(9) 2 626 − 1 251
(10) 3 714 − 561	(11) 2 906 − 2 324	(12) 4 834 − 3 573
(13) 7 231 − 5 410	(14) 2 337 − 1 616	(15) 4 158 − 3 926
(16) 4 077 − 1 824	(17) 1 086 − 741	(18) 3 256 − 2 078

(19) 4 321 − 1 234   (20) 5 260 − 4 710   (21) 3 475 − 1 296
(22) 3 276 − 1 388   (23) 4 251 − 1 358   (24) 1 906 − 1 199
(25) 8 054 − 3 254   (26) 7 932 − 1 085   (27) 7 052 − 1 092
(28) 5 212 − 4 213   (29) 8 006 − 7 007   (30) 4 793 − 3 794

Find the differences between:

39

(1) 2 278 and 1 056   (2) 3 524 and 4 879   (3) 1 096 and 89
(4) 542 and 3 376   (5) 5 234 and 8 016   (6) 785 and 6 155
(7) 4 312 and 6 248   (8) 2 000 and 1 238   (9) 2 573 and 1 564
(10) 6 278 and 7 161   (11) 2 855 and 2 374   (12) 1 025 and 96
(13) 359 and 3 484   (14) 6 278 and 5 118   (15) 279 and 1 814
(16) 4 115 and 413   (17) 3 206 and 1 298   (18) 5 732 and 4 733
(19) Eight thousand seven hundred and seventy-three and Four thousand six hundred and four.
(20) One thousand five hundred and sixty-five and Two thousand six hundred and ninety-one.

PROBLEMS   + −

40

(1) In the picture house; 438 men, 528 women and 96 children were seated. How many tickets were sold?

(2) In building a house; the bricklayers used 657 bricks on one wall, 842 bricks on a second wall, 1 352 bricks on a third wall and 1 185 bricks on the fourth wall. How many bricks were needed?

(3) The gardeners planted 1 200 rose-bushes. If 85 of them died, how many bushes remained?

(4) There were 1 922 people on board a ship. 597 were members of the crew and the rest were passengers. How many passengers were on board?

(5) Mother opened two tins of meat for lunch. One tin held 436 grams of meat and the other held 978 grams of meat. How many grams of meat had we for lunch?

(6) There were 548 pupils on the roll of a school. 29 pupils left and 18 joined the school. How many are now on the roll?

(7) I buy a new car. The car costs £875 plus £43 for extras. The dealer is taking £270 from the price in exchange for my old car. How much do I pay?

Progress Checks

PROGRESS CHECK 1

(1) Copy and write the correct sign for $\triangle$:
$14 + 9 = 20 \triangle 3$

(2) Write this addition sentence as 2 subtraction sentences:
$18 + 15 = 33$

(3) Find $\square$: $24 \div \square = 8$

(4) Write in figures:
two thousand and thirty-four.

(5) Write in symbols: Take 24 from 35 and the result is eleven.

(6) Write the name of this number: 3 201

(7) Find the missing figures:
$82 - ** = 21$
Find the answer to:

(8) $264 + 385 + 749$

(9) $5\,124 + 3\,456 + 1\,289$

(10) $3\,467 - 2\,196$

PROGRESS CHECK 2

(1) Copy and write the correct sign for $\triangle$:
$38 \triangle 26 = 4 \times 3$

(2) Write this division sentence as 2 multiplication sentences:
$84 \div 4 = 21$

(3) Find $\square$: $82 - \square = 50$

(4) Write in figures:
six thousand and seven.

(5) Write in symbols: Five multiplied by 26 equals one hundred and thirty.

(6) Write the name of this number: 5 269

(7) Find the missing figures:
$36 + ** = 89$
Find the answer to:

(8) $526 - 337$

(9) $2\,467 + 1\,593 + 3\,208$

(10) $4\,594 - 3\,197$

PROGRESS CHECK 3

(1) Father has read 183 pages of his book. There are 306 pages in the book. How many pages has he still to read?

(2) There are 267 fruit trees and 425 other trees in a park. How many trees are in the park?

(3) What number must I add to 3 462 to get 6 370?

(4) Mother bought a pen which costs 37p. How much change did she get from a 'Fifty'?

(5) Father planted 87 bulbs into 3 rows. Each row had the same number of bulbs. How many bulbs were in each row?

(6) There are 5 children sitting on each bench. How many children are sitting on 65 benches?

Multiplication of Hundreds

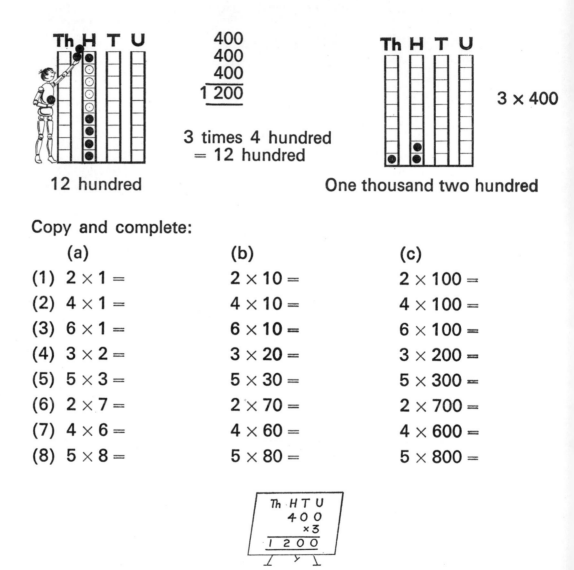

400
400
400
$\overline{1\ 200}$

3 times 4 hundred
= 12 hundred

12 hundred

3×400

One thousand two hundred

Copy and complete:

41

	(a)	(b)	(c)
(1)	$2 \times 1 =$	$2 \times 10 =$	$2 \times 100 =$
(2)	$4 \times 1 =$	$4 \times 10 =$	$4 \times 100 =$
(3)	$6 \times 1 =$	$6 \times 10 =$	$6 \times 100 =$
(4)	$3 \times 2 =$	$3 \times 20 =$	$3 \times 200 =$
(5)	$5 \times 3 =$	$5 \times 30 =$	$5 \times 300 =$
(6)	$2 \times 7 =$	$2 \times 70 =$	$2 \times 700 =$
(7)	$4 \times 6 =$	$4 \times 60 =$	$4 \times 600 =$
(8)	$5 \times 8 =$	$5 \times 80 =$	$5 \times 800 =$

Find the answers to:

42

(1) 2×200	(2) 3×300	(3) 2×400	(4) 2×600
(5) 2×300	(6) 700×3	(7) 800×3	(8) 600×3
(9) 900×4	(10) 700×4	(11) 4×800	(12) 4×600
(13) 5×500	(14) 5×700	(15) 5×900	(16) 600×5
(17) 300×6	(18) 600×6	(19) 900×6	(20) 700×6

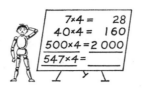

547 × 4

Can you tell Abe the answer?

Copy and complete:

43

(1)
$9 \times 2 = \quad 18$
$50 \times 2 = \quad 100$
$600 \times 2 = 1\,200$
$659 \times 2 = 1\,318$

(2)
$5 \times 3 =$
$40 \times 3 = \quad 120$
$700 \times 3 = 2\,100$
$745 \times 3 =$

(3)
$2 \times 6 = 12$
$80 \times 6 =$
$400 \times 6 =$
$482 \times 6 =$

(4)
$4 \times 2 =$
$50 \times 2 =$
$300 \times 2 =$
$354 \times 2 =$

(5)
$7 \times 5 =$
$90 \times 5 =$
$900 \times 5 =$
$997 \times 5 =$

(6)
$8 \times 3 =$
$70 \times 3 =$
$800 \times 3 =$
$878 \times 3 =$

(7)
$3 \times 4 =$
$60 \times 4 =$
$500 \times 4 =$
$563 \times 4 =$

(8)
$5 \times 6 =$
$200 \times 6 =$
$205 \times 6 =$

(9)
$90 \times 4 =$
$900 \times 4 =$
$990 \times 4 =$

867 × 5

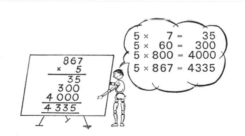

5 × 867

Note: 867 × 5 equals 5 × 867.

Find the products:

44

(1) 134×2 (2) 2×726 (3) 2×683 (4) 975×2

(5) 808×2 (6) 3×312 (7) 468×3 (8) 762×3

(9) 3×534 (10) 650×3 (11) 123×4 (12) 4×512

(13) 941×4 (14) 784×4 (15) 4×607 (16) 105×5

(17) 461×5 (18) 235×5 (19) 5×664 (20) 5×320

(21) 6×113 (22) 524×6 (23) 6×497 (24) 948×6

(25) 205×6 (26) 864×2 (27) 3×369 (28) 375×4

(29) 600×5 (30) 6×166

PROBLEMS

45

(1) We travelled to the coast in 5 buses. Each bus held 72 children. How many children went?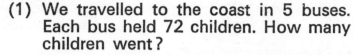

(2) John buys 3 comics each week. There are 52 weeks in a year. How many comics does he buy in a year?

(3) A box holds 144 sticks of chalk. How many sticks of chalk are there in 2 boxes?

(4) Abe's puzzle:  I eat 4 meals a day. There are 365 days in a year. How many meals do I eat in a year?

(5) What is the product of 369 and 3?

(6) A bottle of fruit squash makes 17 drinks. How many drinks can be made from 6 bottles of squash?

(7) A box holds 76 matches. How many matches are there in 5 full boxes?

(8) Each classroom has 48 desks. How many desks are there in 6 classrooms?

(9) A lorry driver packs 470 bags on his lorry each trip. How many bags does he carry in 4 trips?

(10) Find the answer to: $1\,000 - (328 \times 2)$.

Division with Hundreds

$4 \times 30 = 120$
can be written
$120 \div 4 = 30$

Abe writes:

$4 \times 30 = 120$
$$\begin{array}{r} 30 \\ 4\overline{)120} \\ \underline{120} \end{array}$$

46 Write these products as divisions like Abe:
(1) $2 \times 90 = 180$
(2) $3 \times 91 = 273$
(3) $4 \times 72 = 288$
(4) $5 \times 80 = 400$
(5) $6 \times 50 = 300$
(6) $2 \times 94 = 188$
(7) $3 \times 42 = 126$
(8) $4 \times 82 = 328$
(9) $5 \times 71 = 355$
(10) $6 \times 71 = 426$

30 tens ÷ 6 = 5 tens
6 units ÷ 6 = 1 unit
306 ÷ 6 = 51

$$\begin{array}{r} H\ T\ U \\ 5\ 1 \\ 6\overline{)3\ 0\ 6} \\ 3\ 0\ 0 \\ \hline 6 \\ \underline{6} \end{array}$$

$306 \div 6$

Division:

47
(1) $140 \div 2$
(2) $126 \div 2$
(3) $108 \div 2$
(4) $153 \div 3$
(5) $210 \div 3$
(6) $183 \div 3$
(7) $160 \div 4$
(8) $240 \div 4$
(9) $204 \div 4$
(10) $150 \div 5$
(11) $250 \div 5$
(12) $455 \div 5$
(13) $120 \div 6$
(14) $540 \div 6$
(15) $246 \div 6$
(16) $166 \div 2$
(17) $249 \div 3$
(18) $200 \div 4$
(19) $305 \div 5$
(20) $186 \div 6$

485 equals 48 tens
and 5 units

$$\begin{array}{r} H\ T\ U \\ 9\ 7 \\ 5\overline{)4\ 8\ 5} \\ 4\ 5\ 0 \\ \hline 3\ 5 \\ \underline{3\ 5} \end{array}$$

$485 \div 5$

Find the quotients:

48
(1) $2\overline{)146}$
(2) $2\overline{)172}$
(3) $2\overline{)138}$
(4) $2\overline{)150}$
(5) $2\overline{)114}$
(6) $3\overline{)129}$
(7) $3\overline{)165}$
(8) $3\overline{)261}$
(9) $3\overline{)207}$
(10) $3\overline{)234}$

Find the answers to:

(11) 148 ÷ 4   (12) 172 ÷ 4   (13) 236 ÷ 4   (14) 304 ÷ 4

(15) 312 ÷ 4   (16) 125 ÷ 5   (17) 185 ÷ 5   (18) 215 ÷ 5

(19) 340 ÷ 5   (20) 470 ÷ 5   (21) 132 ÷ 6   (22) 264 ÷ 6

(23) 336 ÷ 6   (24) 468 ÷ 6   (25) 534 ÷ 6   (26) 196 ÷ 2

(27) 372 ÷ 4   (28) 295 ÷ 5   (29) 285 ÷ 3   (30) 396 ÷ 6

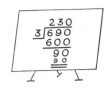

690 ÷ 3

Find the answers to:

49

(1) 2⟌480   (2) 3⟌360   (3) 4⟌840   (4) 5⟌555   (5) 6⟌606

(6) 842 ÷ 2   (7) 936 ÷ 3   (8) 496 ÷ 4   (9) 585 ÷ 5

(10) 672 ÷ 6   (11) 478 ÷ 2   (12) 975 ÷ 3   (13) 672 ÷ 4

(14) 620 ÷ 5   (15) 906 ÷ 6   (16) 584 ÷ 2   (17) 842 ÷ 3

(18) 732 ÷ 4   (19) 825 ÷ 5   (20) 942 ÷ 6   (21) 904 ÷ 4

(22) 314 ÷ 2   (23) 995 ÷ 5   (24) 720 ÷ 6   (25) 525 ÷ 3

Remainders

976 ÷ 6

Find the quotients and remainders:

50

(1) 157 ÷ 2   (2) 243 ÷ 2   (3) 359 ÷ 2   (4) 455 ÷ 2

(5) 561 ÷ 2   (6) 428 ÷ 3   (7) 538 ÷ 3   (8) 761 ÷ 3

(9) 782 ÷ 3   (10) 893 ÷ 3   (11) 557 ÷ 4   (12) 746 ÷ 4

(13) 715 ÷ 4   (14) 905 ÷ 4   (15) 943 ÷ 4   (16) 662 ÷ 5

(17) 674 ÷ 5   (18) 723 ÷ 5   (19) 856 ÷ 5   (20) 988 ÷ 5

(21) 634 ÷ 6   (22) 675 ÷ 6   (23) 796 ÷ 6   (24) 823 ÷ 6

PROBLEMS ÷

51

(1) There are 486 pupils in a school. Half of them are boys. How many boys are there?

(2) A baker made 222 cakes. He put 6 in each box. How many boxes can he fill?

(3) Find one-quarter of 3 528.

(4) A school has 486 pupils. They sit 6 to a bench in the hall. How many benches are needed?

(5) A car travels 415 kilometres in 5 hours. How many kilometres should it travel in 1 hour at the same speed?

(6) A cook fried 420 sausages. Each person had to get 3 sausages. How many dinners did the cook prepare?

(7) Abe's problem:

I counted 360 fingers in the room. Each hand has 5 fingers. How many hands are there?

(8) Use your answer to Abe's problem to find how many people are in the room.

(9) The school library has 476 books shared equally in 4 cases. How many books are in each case?

(10) Three men shared a prize of £540. How many pounds should each man get?

Sevens

7×7

52 Find □:

(1) $1 \times 7 = \square$ (2) $2 \times 7 = \square$ (3) $3 \times 7 = \square$

(4) $4 \times 7 = \square$ (5) $5 \times 7 = \square$ (6) $6 \times 7 = \square$

Copy and fill in the missing numbers:

(7) 0, 7, 14, 21, *, 35, 42, *, *, *, 70.

53 Find □:   (1) $7 \times 0 = \square$ (2) $7 \times 1 = \square$ (3) $7 \times 2 = \square$

(4) $7 \times 3 = \square$ (5) $7 \times 4 = \square$ (6) $7 \times 5 = \square$ (7) $7 \times 6 = \square$

(8) $7 \times 7 = \square$ (9) $7 \times 8 = \square$ (10) $7 \times 9 = \square$ (11) $7 \times 10 = \square$

Check your answers with this table:

X	0	1	2	3	4	5	6	7	8	9	10
1	0	1	2	3	4	5	6	7	8	9	10
2	0	2	4	6	8	10	12	14	16	18	20
3	0	3	6	9	12	15	18	21	24	27	30
4	0	4	8	12	16	20	24	28	32	36	40
5	0	5	10	15	20	25	30	35	40	45	50
6	0	6	12	18	24	30	36	42	48	54	60
7	0	7	14	21	28	35	42	49	56	63	70

54 Multiply by 7:   (1)   2   (2)   4   (3)   5

(4)   1   (5)   3   (6)   7   (7)   9

(8)   6   (9)   0   (10)   8   (11) 10

Find the answers to:

(12)   7 'Tens'   (13)   7 'Fives'

How many new pence?   How many new pence?

There are 7 days in a week.

(14) How many days in 4 weeks?

(15) How many days in 8 weeks?

(16) How many days in 6 weeks?

Find the answers to:

55

(1) 7×2 (2) 4×7 (3) 7×9 (4) 3×7 (5) 8×7

(6) 7×7 (7) 2×7 (8) 7×1 (9) 7×4 (10) 7×3

(11) 6×7 (12) 5×7 (13) 1×7 (14) 10×7 (15) 7×5

(16) 7×6 (17) 7×10 (18) 9×7 (19) 0×7 (20) 7×8

(21) 11×7 (22) 13×7 (23) 7×12 (24) 7×11 (25) 14×7

Find the value of:

56

(1) $(7 \times 5) + 2$ (2) $(7 \times 8) + 5$ (3) $(3 \times 7) + 7$

(4) $(2 \times 7) + 1$ (5) $(7 \times 9) + 4$ (6) $(7 \times 4) + 3$

(7) $(7 \times 6) + 5$ (8) $(7 \times 1) + 3$ (9) $(7 \times 10) + 6$

(10) $(7 \times 7) + 2$ (11) $(7 \times 0) + 8$ (12) $(6 \times 7) + 8$

(13) $(5 \times 7) - 3$ (14) $(9 \times 7) - 3$ (15) $(7 \times 2) - 8$

(16) $(10 \times 7) - 5$ (17) $(7 \times 3) - 6$ (18) $(8 \times 7) - 9$

(19) $(7 \times 6) - 7$ (20) $(4 \times 7) + 5$

30×7 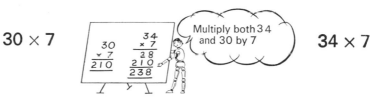 34×7

Multiplication:

57

(1) 20×7 (2) 40×7 (3) 60×7 (4) 50×7 (5) 90×7

(6) 70×7 (7) 80×7 (8) 31×7 (9) 21×7 (10) 61×7

(11) 22×7 (12) 32×7 (13) 53×7 (14) 43×7 (15) 64×7

(16) 74×7 (17) 85×7 (18) 96×7 (19) 27×7 (20) 58×7

(21) 39×7 (22) 44×7 (23) 16×7 (24) 83×7 (25) 76×7

Abe notes that teacher does not always write the working figures. He wants to try this.

54 × 7

Find the answers to:

58

(1) 31 × 7	(2) 21 × 7	(3) 61 × 7	(4) 51 × 7	(5) 91 × 7
(6) 81 × 7	(7) 71 × 7	(8) 32 × 7	(9) 52 × 7	(10) 63 × 7
(11) 94 × 7	(12) 35 × 7	(13) 65 × 7	(14) 86 × 7	(15) 68 × 7
(16) 97 × 7	(17) 49 × 7	(18) 78 × 7	(19) 15 × 7	(20) 99 × 7

With Working

```
  238
×   7
   56
  210
 1400
 1666
```

Without Working

```
  238
×   7
  2 5
 1666
```

238 × 7

Find the products, with working:

59

(1) 124 × 7	(2) 127 × 7	(3) 134 × 7	(4) 136 × 7
(5) 139 × 7	(6) 140 × 7	(7) 105 × 7	(8) 156 × 7
(9) 245 × 7	(10) 268 × 7	(11) 283 × 7	(12) 287 × 7
(13) 320 × 7	(14) 372 × 7	(15) 574 × 7	(16) 628 × 7
(17) 747 × 7	(18) 860 × 7	(19) 453 × 7	(20) 179 × 7

Find these products without working:

(21) 410 × 7	(22) 121 × 7	(23) 108 × 7	(24) 120 × 7
(25) 304 × 7	(26) 381 × 7	(27) 532 × 7	(28) 785 × 7
(29) 907 × 7	(30) 999 × 7		

PROBLEMS

60

(1) Mother gets 3 bottles of milk each day. How many bottles does she get in a week?

(2) James wants to buy a set of 7 model cars. They cost 9p each. How much must he save?

(3) Dad says he drinks 8 cups of tea each day. How many cups of tea does he drink in a week?

(4) I can read 7 pages of my book each night. How many pages can I read in a week?

(5) Abe's problem:

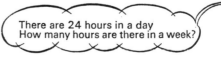

There are 24 hours in a day
How many hours are there in a week?

NUMBER SENTENCES

61 Find □ :

(1) $\square \times 7 = 35$ (2) $7 \times \square = 14$ (3) $7 \times \square = 56$
(4) $\square \times 7 = 21$ (5) $7 \times \square = 49$ (6) $7 \times \square = 28$
(7) $7 \times \square = 7$ (8) $\square \times 7 = 70$ (9) $\square \times 7 = 63$
(10) $7 \times \square = 42$ (11) $\square \times 7 = 0$ (12) $\square \times 7 = 56$

DIVISION BY SEVEN   ÷ 7

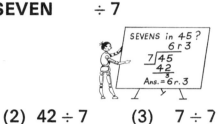

Division

62

(1) $28 \div 7$ (2) $42 \div 7$ (3) $7 \div 7$ (4) $21 \div 7$
(5) $35 \div 7$ (6) $14 \div 7$ (7) $63 \div 7$ (8) $49 \div 7$
(9) $0 \div 7$ (10) $56 \div 7$ (11) $70 \div 7$ (12) $16 \div 7$
(13) $25 \div 7$ (14) $34 \div 7$ (15) $36 \div 7$ (16) $46 \div 7$
(17) $50 \div 7$ (18) $58 \div 7$ (19) $66 \div 7$ (20) $72 \div 7$
(21) $10 \div 7$ (22) $3 \div 7$ (23) $41 \div 7$ (24) $30 \div 7$
(25) $19 \div 7$ (26) $54 \div 7$ (27) $65 \div 7$ (28) $59 \div 7$
(29) $24 \div 7$ (30) $9 \div 7$

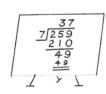

$259 \div 7$

Find the answers to:

63

(1) $7\overline{)77}$ (2) $7\overline{)147}$ (3) $7\overline{)210}$ (4) $7\overline{)287}$ (5) $7\overline{)420}$
(6) $357 \div 7$ (7) $707 \div 7$ (8) $490 \div 7$ (9) $567 \div 7$
(10) $637 \div 7$ (11) $168 \div 7$ (12) $224 \div 7$ (13) $441 \div 7$
(14) $84 \div 7$ (15) $189 \div 7$ (16) $98 \div 7$ (17) $112 \div 7$
(18) $231 \div 7$ (19) $91 \div 7$ (20) $168 \div 7$ (21) $315 \div 7$
(22) $133 \div 7$ (23) $266 \div 7$ (24) $392 \div 7$ (25) $469 \div 7$
(26) $539 \div 7$ (27) $602 \div 7$ (28) $665 \div 7$ (29) $798 \div 7$

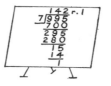

995 ÷ 7

Find the quotients and remainders:

64

(1) 7 | 85 (2) 7 | 107 (3) 7 | 133 (4) 7 | 186

(5) 7 | 229 (6) 7 | 265 (7) 7 | 297 (8) 7 | 323

(9) 408 ÷ 7 (10) 431 ÷ 7 (11) 453 ÷ 7 (12) 487 ÷ 7

(13) 520 ÷ 7 (14) 547 ÷ 7 (15) 598 ÷ 7 (16) 632 ÷ 7

(17) 690 ÷ 7 (18) 724 ÷ 7 (19) 766 ÷ 7 (20) 811 ÷ 7

(21) 835 ÷ 7 (22) 873 ÷ 7 (23) 960 ÷ 7

PROBLEMS

65

(1) A grocer took 42 eggs out of a case and put them in 7 small boxes. How many eggs does each box hold?

(2) 7 bottles of lemonade cost 56p. Find the cost of 1 bottle.

(3) Find one-seventh of 63.

(4) How many weeks are there in 49 days?

(5) The chemist counted out 252 tablets equally into 7 bottles. How many did he put into each bottle?

(6) The parents sat 7 on a bench at the school concert. There were 294 parents at the concert. How many benches were needed?

(7) I have 28 meals each week in my house. How many meals do I have each day?

(8) Seven bakers have to make 406 cakes. How many should each make?

(9) If 7 buses can carry 504 passengers, how many can each bus carry?

(10) A machine turns out 392 model boats in 7 hours. How many boats can it turn out in an hour?

Eights

8×8

66 Find □ :
   (1) $1 \times 8 = □$ (2) $2 \times 8 = □$ (3) $3 \times 8 = □$
   (4) $4 \times 8 = □$ (5) $5 \times 8 = □$ (6) $6 \times 8 = □$
Copy and fill in the missing figures:
   (7) 0, 8, 16, 24, *, 40, 48, *, *, *, 80

67 Find □ :
                  (1) $8 \times 0 = □$ (2) $8 \times 1 = □$ (3) $8 \times 2 = □$
  (4) $8 \times 3 = □$ (5) $8 \times 4 = □$ (6) $8 \times 5 = □$ (7) $8 \times 6 = □$
  (8) $8 \times 7 = □$ (9) $8 \times 8 = □$ (10) $8 \times 9 = □$ (11) $8 \times 10 = □$
Check your answers with the multiplication table on **Page 48**.

68 Multiply by 8:
            (1) 4   (2) 3   (3) 2   (4) 1   (5) 5
  (6) 10   (7) 0   (8) 7   (9) 6   (10) 9   (11) 8
Find the answers to

(12)   8 'Tens'   (13) 8 'Fives'
                  How many   How many
                  new pence?   new pence?

69 Find the answers to:
   (1) 8×2 (2) 8×4 (3) 1×8 (4) 8×8 (5) 8×3
   (6) 7×8 (7) 5×8 (8) 8×6 (9) 10×8 (10) 0×8
 (11) 9×8 (12) 2×8 (13) 3×8 (14) 6×8 (15) 8×7
 (16) 8×10 (17) 8×5 (18) 2×8 (19) 8×9 (20) 4×8

70 Find the value of:
   (1) $(8 \times 2) + 2$ (2) $(8 \times 5) + 3$ (3) $(8 \times 1) + 4$ (4) $(8 \times 6) + 2$
   (5) $(8 \times 10) + 6$ (6) $(8 \times 7) + 4$ (7) $(8 \times 9) + 3$ (8) $(8 \times 4) + 6$
   (9) $(8 \times 3) + 2$ (10) $(8 \times 8) + 7$ (11) $(4 \times 8) + 9$ (12) $(8 \times 0) + 5$
 (13) $(8 \times 2) - 4$ (14) $(8 \times 7) - 5$ (15) $(8 \times 3) - 4$ (16) $(5 \times 8) - 5$
 (17) $(8 \times 8) - 4$ (18) $(9 \times 8) - 3$ (19) $(6 \times 8) - 5$ (20) $(3 \times 8) - 7$

PROBLEMS

71

(1) Mother bought 6 packets of paper cups for the party. There are 8 cups in a packet. How many cups did she buy?

(2) What is the product of 8 and 9?

(3) A packet of biscuits costs 8p. What will 8 packets cost?

(4)  A bar of toffee has 8 pieces. How many pieces have 4 bars of toffee?

(5) A litre bottle will fill 8 cups. How many cups will 7 litre bottles fill?

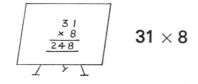

31×8

Find the answers:

72

(1) 20×8 (2) 60×8 (3) 30×8 (4) 50×8 (5) 80×8
(6) 11×8 (7) 51×8 (8) 71×8 (9) 21×8 (10) 81×8
(11) 70×8 (12) 90×8 (13) 61×8 (14) 41×8 (15) 91×8

With working	Without working
56 × 8 48 400 448	56 × 8 448

Find the answers with working:

73

(1) 14×8 (2) 25×8 (3) 28×8 (4) 45×8 (5) 46×8
(6) 55×8 (7) 57×8 (8) 66×8 (9) 82×8 (10) 87×8

Find the answers without the working figures:

(11) 17×8 (12) 19×8 (13) 22×8 (14) 33×8 (15) 36×8
(16) 38×8 (17) 47×8 (18) 49×8 (19) 54×8 (20) 59×8
(21) 63×8 (22) 64×8 (23) 68×8 (24) 72×8 (25) 74×8
(26) 76×8 (27) 89×8 (28) 93×8 (29) 95×8 (30) 98×8

754×8

Multiplication:

74

(1) 123×8 (2) 128×8 (3) 132×8 (4) 266×8
(5) 264×8 (6) 223×8 (7) 355×8 (8) 367×8

(9) 439×8 (10) 471×8 (11) 548×8 (12) 582×8
(13) 656×8 (14) 694×8 (15) 743×8 (16) 770×8
(17) 831×8 (18) 899×8 (19) 957×8 (20) 985×8

Now try these without showing the working figures:

(21) 700×8 (22) 402×8 (23) 508×8 (24) 806×8
(25) 750×8 (26) 220×8 (27) 350×8 (28) 113×8
(29) 263×8 (30) 999×8

NUMBER SENTENCES

Find □ :

75

(1) $8 \times \square = 32$ (2) $8 \times \square = 48$ (3) $\square \times 8 = 24$
(4) $\square \times 8 = 16$ (5) $\square \times 8 = 56$ (6) $8 \times \square = 8$
(7) $8 \times \square = 72$ (8) $\square \times 8 = 40$ (9) $8 \times \square = 64$
(10) $\square \times 8 = 80$

DIVISION BY EIGHT $\div 8$

$$50 \div 8$$

Find the answers:

76

(1) $16 \div 8$ (2) $72 \div 8$ (3) $24 \div 8$ (4) $56 \div 8$ (5) $48 \div 8$
(6) $32 \div 8$ (7) $64 : 8$ (8) $80 \div 8$ (9) $40 \div 8$ (10) $0 \div 8$
(11) $8 \div 8$ (12) $33 \div 8$ (13) $43 \div 8$ (14) $17 \div 8$ (15) $50 \div 8$
(16) $26 \div 8$ (17) $60 \div 8$ (18) $69 \div 8$ (19) $73 \div 8$ (20) $81 \div 8$
(21) $9 \div 8$ (22) $15 \div 8$ (23) $29 \div 8$ (24) $39 \div 8$ (25) $47 \div 8$
(26) $55 \div 8$ (27) $70 \div 8$ (28) $6 \div 8$ (29) $19 \div 8$ (30) $85 \div 8$

$$
\begin{array}{r}
117 \\
8\overline{)936} \\
800 \\
\hline
136 \\
80 \\
\hline
56 \\
\end{array}
$$

$$936 \div 8$$

Division:

77

(1) $8\overline{)88}$ (2) $8\overline{)96}$ (3) $8\overline{)152}$ (4) $8\overline{)728}$ (5) $8\overline{)896}$
(6) $8\overline{)208}$ (7) $8\overline{)392}$ (8) $8\overline{)408}$ (9) $8\overline{)456}$ (10) $8\overline{)664}$
(11) $960 \div 8$ (12) $168 \div 8$ (13) $224 \div 8$ (14) $264 \div 8$
(15) $296 \div 8$ (16) $336 \div 8$ (17) $376 \div 8$ (18) $440 \div 8$
(19) $512 \div 8$ (20) $608 \div 8$ (21) $640 \div 8$ (22) $680 \div 8$
(23) $744 \div 8$ (24) $768 \div 8$ (25) $808 \div 8$ (26) $912 \div 8$
(27) $968 \div 8$ (28) $320 \div 8$ (29) $840 \div 8$ (30) $872 \div 8$

Find the quotients and remainders:

78

(1) 8⟌90   (2) 8⟌97   (3) 8⟌105   (4) 8⟌153   (5) 8⟌186

(6) 8⟌210   (7) 8⟌235   (8) 8⟌258   (9) 8⟌283   (10) 8⟌331

Find the answers to

(11) 356 ÷ 8   (12) 389 ÷ 8   (13) 404 ÷ 8   (14) 461 ÷ 8

(15) 477 ÷ 8   (16) 497 ÷ 8   (17) 540 ÷ 8   (18) 566 ÷ 8

(19) 679 ÷ 8   (20) 691 ÷ 8   (21) 897 ÷ 8   (22) 988 ÷ 8

(23) 839 ÷ 8   (24) 878 ÷ 8   (25) 850 ÷ 8

PROBLEMS   ÷ 8

79

(1) Dad needs 48 screws. There are 8 in each packet. How many packets must he buy?

(2) The gardener at the park plants 72 young trees in 8 rows. How many does he plant in each row?

(3) I can save 8p each week. How many weeks will it take me to save 64p?

(4) Divide 96 by 8.

(5) The grocer put 104 cans of fruit in stacks of eight. How many stacks did he have?

(6) The janitor needs 8 light bulbs each week. He has 60. How many weeks will they last?

(7) A packet holds 8 biscuits. How many packets are needed to hold 512 biscuits?

(8) A train travels 664 kilometres in 8 hours. How many kilometres does it travel in 1 hour at the same speed?

(9) There were 56 cakes equally shared in 8 plates. How many were on each plate?

(10) There are more than 30 and less than 40 in a class. The teacher can arrange the pupils in groups of 8. How many are in the class?

Eighths $\frac{1}{8}$

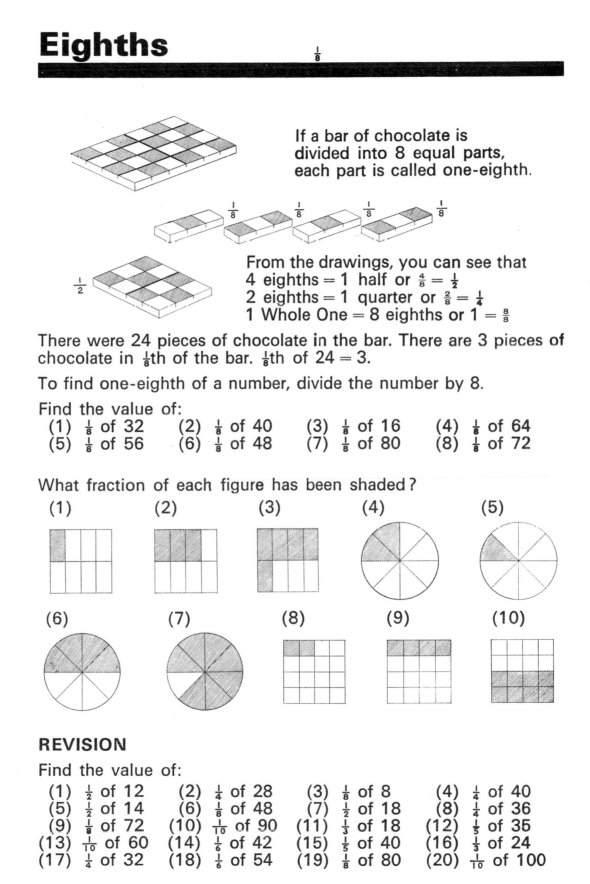

If a bar of chocolate is divided into 8 equal parts, each part is called one-eighth.

From the drawings, you can see that

4 eighths = 1 half or $\frac{4}{8} = \frac{1}{2}$

2 eighths = 1 quarter or $\frac{2}{8} = \frac{1}{4}$

1 Whole One = 8 eighths or $1 = \frac{8}{8}$

There were 24 pieces of chocolate in the bar. There are 3 pieces of chocolate in $\frac{1}{8}$th of the bar. $\frac{1}{8}$th of 24 = 3.

To find one-eighth of a number, divide the number by 8.

Find the value of:

80

(1) $\frac{1}{8}$ of 32   (2) $\frac{1}{8}$ of 40   (3) $\frac{1}{8}$ of 16   (4) $\frac{1}{8}$ of 64

(5) $\frac{1}{8}$ of 56   (6) $\frac{1}{8}$ of 48   (7) $\frac{1}{8}$ of 80   (8) $\frac{1}{8}$ of 72

What fraction of each figure has been shaded?

(1)   (2)   (3)   (4)   (5)

(6)   (7)   (8)   (9)   (10)

REVISION

Find the value of:

81

(1) $\frac{1}{2}$ of 12   (2) $\frac{1}{4}$ of 28   (3) $\frac{1}{8}$ of 8   (4) $\frac{1}{4}$ of 40

(5) $\frac{1}{2}$ of 14   (6) $\frac{1}{8}$ of 48   (7) $\frac{1}{2}$ of 18   (8) $\frac{1}{4}$ of 36

(9) $\frac{1}{8}$ of 72   (10) $\frac{1}{10}$ of 90   (11) $\frac{1}{3}$ of 18   (12) $\frac{1}{5}$ of 35

(13) $\frac{1}{10}$ of 60   (14) $\frac{1}{6}$ of 42   (15) $\frac{1}{5}$ of 40   (16) $\frac{1}{3}$ of 24

(17) $\frac{1}{4}$ of 32   (18) $\frac{1}{6}$ of 54   (19) $\frac{1}{8}$ of 80   (20) $\frac{1}{10}$ of 100

Nines

9×9

82

Find □:
(1) $1 \times 9 = \square$ (2) $2 \times 9 = \square$ (3) $3 \times 9 = \square$
(4) $4 \times 9 = \square$ (5) $5 \times 9 = \square$ (6) $6 \times 9 = \square$

Copy and fill in the missing figures:
(7) 0, 9, 18, 27, *, 45, 54, *, *, *, 90

83

Find □:   (1) $9 \times 0 = \square$ (2) $9 \times 1 = \square$ (3) $9 \times 2 = \square$
(4) $9 \times 3 = \square$ (5) $9 \times 4 = \square$ (6) $9 \times 5 = \square$ (7) $9 \times 6 = \square$
(8) $9 \times 7 = \square$ (9) $9 \times 8 = \square$ (10) $9 \times 9 = \square$ (11) $9 \times 10 = \square$

Check your answers with the table on Page 48.

84

Multiply by 9:
          (1) 3   (2) 5   (3) 4   (4) 1   (5) 7
(6) 2   (7) 8   (8) 6   (9) 10   (10) 0   (11) 9

Find the answers to:

(12)   9 'Tens'
                            How many
                            new pence?

(13)   9 'Fives'
                            How many
                            new pence?

85

Find the answers to:
(1) 9×2 (2) 1×9 (3) 9×4 (4) 3×9 (5) 9×6
(6) 9×9 (7) 10×9 (8) 7×9 (9) 8×9 (10) 9×5
(11) 9×1 (12) 9×10 (13) 0×9 (14) 9×3 (15) 9×7
(16) 4×9 (17) 9×8 (18) 6×9 (19) 2×9 (20) 5×9

86

What is the value of:
(1) $(9 \times 5) + 3$ (2) $(9 \times 8) + 5$ (3) $(9 \times 3) + 3$ (4) $(2 \times 9) + 2$
(5) $(7 \times 9) + 7$ (6) $(9 \times 4) + 6$ (7) $(9 \times 6) + 7$ (8) $(1 \times 9) + 8$
(9) $(10 \times 9) + 10$ (10) $(9 \times 9) + 1$ (11) $(0 \times 9) + 6$ (12) $(6 \times 9) - 3$
(13) $(9 \times 5) - 4$ (14) $(7 \times 9) - 3$ (15) $(9 \times 2) - 8$ (16) $(9 \times 10) - 4$
(17) $(3 \times 9) - 8$ (18) $(8 \times 9) - 6$ (19) $(6 \times 9) - 7$ (20) $(4 \times 9) - 7$

PROBLEMS

87

(1) Jack is 9 years old. His father is four times as old. How old is Jack's father?

(2) Mother bought 9 small tins of meat at 9p each. Find the total cost.

(3) Each of us has ten fingers. How many fingers have 9 girls?

(4) There are 8 pieces in a bar of toffee. How many pieces are there in 9 bars of toffee?

(5) How many days are there in 9 weeks and 2 days?

9×61

61×9

Remember $9 \times 61 = 61 \times 9$

Find the answers to:

88

(1) 20×9 (2) 70×9 (3) 30×9 (4) 30×9 (5) 40×9
(6) 9×91 (7) 9×21 (8) 9×51 (9) 9×31 (10) 9×71
(11) 81×9 (12) 11×9 (13) 9×41 (14) 9×50 (15) 50×9

9×64

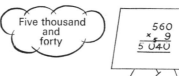

64×9

Multiplication:

89

(1) 13×9 (2) 15×9 (3) 17×9 (4) 22×9 (5) 24×9
(6) 26×9 (7) 29×9 (8) 32×9 (9) 35×9 (10) 38×9
(11) 9×44 (12) 9×47 (13) 9×49 (14) 9×52 (15) 9×53
(16) 9×56 (17) 9×62 (18) 9×65 (19) 9×68 (20) 9×72
(21) 74×9 (22) 76×9 (23) 77×9 (24) 84×9 (25) 87×9
(26) 89×9 (27) 93×9 (28) 95×9 (29) 96×9 (30) 98×9

9×560

Five thousand and forty

560×9

Find the products:

90

(1) 111×9 (2) 9×232 (3) 9×287 (4) 142×9
(5) 346×9 (6) 9×376 (7) 9×453 (8) 491×9
(9) 519×9 (10) 9×564 (11) 9×625 (12) 684×9
(13) 718×9 (14) 9×773 (15) 835×9 (16) 106×9
(17) 9×880 (18) 906×9 (19) 920×9 (20) 970×9
(21) 497×9 (22) 288×9 (23) 9×781 (24) 9×629
(25) 162×9 (26) 9×973 (27) 504×9 (28) 355×9
(29) 236×9 (30) 9×947

NUMBER SENTENCES

Find □:

91
(1) $9 \times \square = 27$ (2) $\square \times 9 = 45$ (3) $9 \times \square = 36$
(4) $\square \times 9 = 18$ (5) $\square \times 9 = 63$ (6) $9 \times \square = 81$
(7) $\square \times 9 = 9$ (8) $9 \times \square = 54$ (9) $\square \times 9 = 72$
(10) $9 \times \square = 0$

DIVISION BY NINE   ÷ 9

Find the answers to:

92
(1) $36 \div 9$ (2) $63 \div 9$ (3) $45 \div 9$ (4) $54 \div 9$ (5) $27 \div 9$
(6) $72 \div 9$ (7) $9 \div 9$ (8) $81 \div 9$ (9) $0 \div 9$ (10) $18 \div 9$
(11) $80 \div 9$ (12) $28 \div 9$ (13) $33 \div 9$ (14) $49 \div 9$ (15) $74 \div 9$
(16) $19 \div 9$ (17) $58 \div 9$ (18) $17 \div 9$ (19) $43 \div 9$ (20) $68 \div 9$
(21) $82 \div 9$ (22) $55 \div 9$ (23) $41 \div 9$ (24) $23 \div 9$ (25) $30 \div 9$
(26) $79 \div 9$ (27) $5 \div 9$ (28) $15 \div 9$ (29) $70 \div 9$ (30) $60 \div 9$

$$936 \div 9$$

With working   Without working

Division:

93
(1) $9\overline{)99}$ (2) $9\overline{)108}$ (3) $9\overline{)189}$ (4) $9\overline{)279}$

(5) $9\overline{)549}$ (6) $9\overline{)720}$ (7) $9\overline{)819}$ (8) $9\overline{)126}$

(9) $9\overline{)171}$ (10) $9\overline{)234}$

Find the answers to:

(11) $288 \div 9$ (12) $342 \div 9$ (13) $396 \div 9$ (14) $450 \div 9$
(15) $504 \div 9$ (16) $612 \div 9$ (17) $666 \div 9$ (18) $702 \div 9$
(19) $828 \div 9$ (20) $873 \div 9$ (21) $936 \div 9$ (22) $423 \div 9$
(23) $531 \div 9$ (24) $648 \div 9$ (25) $756 \div 9$ (26) $864 \div 9$
(27) $972 \div 9$ (28) $315 \div 9$ (29) $459 \div 9$ (30) $675 \div 9$

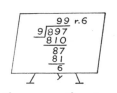

$$897 \div 9$$

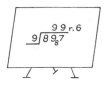

Find the quotients and remainders:

94
(1) $127 \div 9$ (2) $154 \div 9$ (3) $182 \div 9$ (4) $217 \div 9$
(5) $245 \div 9$ (6) $273 \div 9$ (7) $201 \div 9$ (8) $335 \div 9$
(9) $363 \div 9$ (10) $391 \div 9$ (11) $401 \div 9$ (12) $435 \div 9$

(13) 473 ÷ 9	(14) 517 ÷ 9	(15) 545 ÷ 9	(16) 583 ÷ 9
(17) 627 ÷ 9	(18) 656 ÷ 9	(19) 692 ÷ 9	(20) 763 ÷ 9
(21) 741 ÷ 9	(22) 753 ÷ 9	(23) 824 ÷ 9	(24) 830 ÷ 9

PROBLEMS

95

(1) There are 9 tables for 54 pupils at the dining centre. How many should sit at each table?

(2) Find one-ninth of 81.

(3) The baker made 90 cakes and put 9 in each box. How many boxes did he fill?

(4) How many 9p packets of sweets can you buy with a 'Fifty'. How much change will you get?

(5) If 9 buses can carry 432 passengers, how many passengers can each bus take?

(6) The butcher made 279 sausages. He wrapped them up in bundles of 9. How many bundles did he have?

(7) Andrew bought 9 packets of stamps. There were 144 altogether. How many were in each packet?

(8) Add 427 and 104 and divide your answer by 9.

(9) Nine bicycles cost £225. What is the cost of each?

(10) Nine girls decide to make 180 soft toys for a children's hospital. How many should each make?

The Multiplication Square

X	0	1	2	3	4	5	6	7	8	9	10
0	0	0	0	0	0	0	0	0	0	0	0
1	0	1	2	3	4	5	6	7	8	9	10
2	0	2	4	6	8	10	12	14	16	18	20
3	0	3	6	9	12	15	18	21	24	27	30
4	0	4	8	12	16	20	24	28	32	36	40
5	0	5	10	15	20	25	30	35	40	45	50
6	0	6	12	18	24	30	36	42	48	54	60
7	0	7	14	21	28	35	42	49	56	63	70
8	0	8	16	24	32	40	48	56	64	72	80
9	0	9	18	27	36	45	54	63	72	81	90
10	0	10	20	30	40	50	60	70	80	90	100

We can use this table to multiply and to divide.

For examples: $7 \times 4 = 28$ and $4 \times 7 = 28$
$54 \div 6 = 9$ and $54 \div 9 = 6$

96

(1) Copy and complete this diagonal line of numbers from the table:
1, 4, 9, 16, *, *, *, 64, *, 100.

Here are some of these numbers in dot drawings. Which numbers
are they?

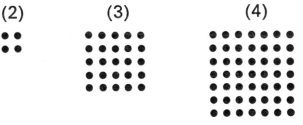

(2)   (3)   (4)

These numbers are called SQUARE NUMBERS.

Draw dot drawings to show that these numbers are square numbers:
(5) 9   (6) 16   (7) 36   (8) 64   (9) 81

Progress Checks

PROGRESS CHECK 4

(1) Copy and write the correct sign for $\triangle$:
$42 \triangle 6 = 15 - 8$

(2) Find $\square$:
$(4 \times 9) + 6 = \square$

(3) Write in figures: nine thousand three hundred and one

Find the answer to:

(4) $8\,295 - 4\,436$

(5) $1\,825 + 2\,073 + 3\,657$

(6) 93×5

(7) $198 \div 2$

(8) $(8 \times 5) - 6$

(9) 164×9

(10) $\frac{1}{8}$ of 328

PROGRESS CHECK 5

(1) Write this multiplication sentence as 2 division sentences: $9 \times 8 = 72$

(2) Find $\square$:
$\square \div 7 = 8$

(3) Write the name of this number:
4082

Find the answer to:

(4) $7\,123 - 3\,346$

(5) $2\,068 + 1\,782 + 969$

(6) 3×464

(7) $315 - 7$

(8) $(9 \times 6) + 7$

(9) 86×8

Find the answer and remainder:
(10) $713 \div 9$

PROGRESS CHECK 6

(1) There are 68 houses in each block of new flats. How many houses are there in 8 blocks of flats?

(2) 252 children travel in 7 buses. There are the same number of children in each bus. How many children are in each bus?

(3) Find the product of 216 and 4.

(4) Winston Churchill was born in the year 1874. His last birthday was in the year 1964. What was his age when he died?

(5) Find the value of ($\frac{1}{8}$ of 48) + ($\frac{1}{6}$ of 54).

(6) What is the remainder on dividing 987 by 9?

(7) In a reading book there are 42 lines on each page. How many lines are there in 9 pages?

(8) Find the sum of $2\,436$ and $3\,752$.

(9) The baker baked 312 cakes and packed them 6 to a box. How many boxes did he fill?

(10) There are 365 days in a year. How many days are there in one-fifth of a year?

Fractions

Mother bakes a cake and divides it into **8** equal parts.
We know that each part is one-eighth of the cake.
We write $\frac{1}{8}$

Two pieces of cake are put to the side.
We say "two eighths" or "one quarter" is put to the side.
We write $\frac{2}{8}$ or $\frac{1}{4}$

What fraction remains?
We write $\frac{6}{8}$ or $\frac{3}{4}$

97 Here are some shapes, called RECTANGLES which have been divided into eight equal parts. What fraction of each rectangle has been shaded?

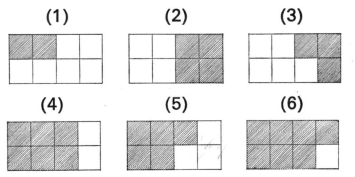

(1) (2) (3)

(4) (5) (6)

Each of these rectangles have been divided into ten equal parts. Write down the fraction which has been shaded:

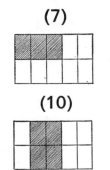

(7) (8) (9)

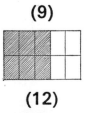

(10) (11) (12)

Each of these rectangles have been divided into 6 equal parts. Write down the fraction which has been shaded:

(13)　　　(14)　　　(15)　　　(16)

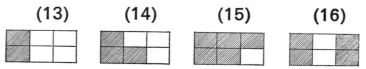

Teacher will give you 4 strips of paper of equal length.

98

(1) On one of the strips, write 1 Whole One.

(2) Take another strip and fold it into 2 equal parts. Write $\frac{1}{2}$ in each of these parts.

(3) Take a third strip and fold it into 2 equal parts. Still folded, fold it again into 2 equal parts. Unfold and you should have 4 equal parts. Write $\frac{1}{4}$ in each of these parts.

(4) Take the last strip and fold it into 4 equal parts. Still folded, fold it again into 2 equal parts. Unfold and you should have 8 equal parts. Write $\frac{1}{8}$ in each of these parts.

(5) With writing showing, place the strips one under each other. You now have a fraction chart.

FRACTION CHART

$1, \frac{1}{2}, \frac{1}{4}, \frac{1}{8}$

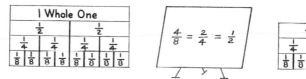

From the fraction chart, copy and complete:

99

(1) $\frac{2}{8} = \frac{}{4}$　　(2) $\frac{6}{8} = \frac{}{4}$　　(3) $\frac{2}{4} = \frac{}{2}$　　(4) $\frac{}{2} = 1$

(5) $\frac{}{4} = 1$　　(6) $\frac{}{8} = 1$　　(7) $\frac{8}{8} = \frac{}{4} = \frac{}{2} =$

ADDITION AND SUBTRACTION OF FRACTIONS

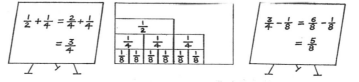

Find the answers to:

100

(1) $\frac{1}{4} + \frac{1}{4}$　　(2) $\frac{3}{4} + \frac{1}{4}$　　(3) $\frac{1}{4} + \frac{1}{4} + \frac{1}{4}$　　(4) $\frac{1}{4} + \frac{1}{2} + \frac{1}{4}$　　(5) $\frac{3}{4} - \frac{1}{2}$

(6) $\frac{3}{4} - \frac{1}{4}$　　(7) $1 - \frac{1}{2}$　　(8) $1 - \frac{3}{4}$　　(9) $\frac{1}{8} + \frac{1}{8}$　　(10) $\frac{1}{4} + \frac{1}{8}$

(11) $\frac{1}{2} + \frac{1}{8}$　　(12) $\frac{3}{4} + \frac{1}{8}$　　(13) $\frac{1}{4} + \frac{3}{8}$　　(14) $\frac{3}{8} + \frac{1}{2}$　　(15) $\frac{1}{4} - \frac{1}{8}$

(16) $\frac{1}{2} - \frac{1}{8}$　　(17) $1 - \frac{1}{8}$　　(18) $1 - \frac{3}{8}$　　(19) $\frac{1}{2} - \frac{3}{8}$　　(20) $\frac{3}{4} - \frac{3}{8}$

FRACTION CHART

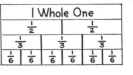 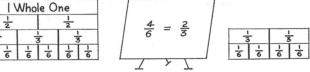

From the fraction chart, copy and complete:

101

(1) $\frac{2}{6} = \frac{}{3}$ (2) $\frac{3}{6} = \frac{}{2}$ (3) $\frac{4}{6} = \frac{}{3}$ (4) $\frac{}{2} = 1$

(5) $\frac{}{3} = 1$ (6) $\frac{}{6} = 1$ (7) $\frac{6}{6} = \frac{}{3} = \frac{}{2} =$

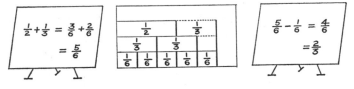

Find the answers to:

102

(1) $\frac{1}{6} + \frac{1}{6}$ (2) $\frac{1}{3} + \frac{1}{6}$ (3) $\frac{1}{2} + \frac{1}{6}$ (4) $\frac{2}{3} + \frac{1}{6}$ (5) $\frac{1}{6} + \frac{5}{6}$

(6) $\frac{1}{3} + \frac{1}{3}$ (7) $\frac{1}{3} + \frac{1}{3}$ (8) $\frac{2}{3} + \frac{1}{3}$ (9) $\frac{1}{2} + \frac{1}{6} + \frac{1}{3}$ (10) $\frac{1}{3} - \frac{1}{6}$

(11) $\frac{1}{2} - \frac{1}{6}$ (12) $\frac{1}{2} - \frac{1}{3}$ (13) $\frac{2}{3} - \frac{1}{6}$ (14) $\frac{2}{3} - \frac{1}{3}$ (15) $\frac{2}{3} - \frac{1}{2}$

(16) $\frac{5}{6} - \frac{1}{3}$ (17) $\frac{5}{6} - \frac{2}{3}$ (18) $1 - \frac{1}{6}$ (19) $1 - \frac{2}{3}$ (20) $1 - \frac{5}{6}$

FRACTION CHART

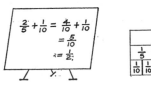

From the fraction chart, copy and complete:

103

(1) $\frac{2}{10} = \frac{}{5}$ (2) $\frac{4}{10} = \frac{}{5}$ (3) $\frac{1}{2} = \frac{}{10}$ (4) $\frac{3}{5} = \frac{}{10}$

(5) $1 = \frac{}{5}$ (6) $1 = \frac{}{10}$ (7) $1 = \frac{}{2} = \frac{}{5} = \frac{}{10}$

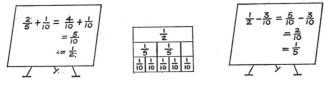

Find the answers to:

104

(1) $\frac{1}{10} + \frac{1}{10}$ (2) $\frac{1}{10} + \frac{3}{10}$ (3) $\frac{7}{10} + \frac{1}{10}$ (4) $\frac{3}{5} + \frac{1}{5}$

(5) $\frac{9}{10} + \frac{1}{10}$ (6) $\frac{3}{10} + \frac{3}{10}$ (7) $\frac{1}{10} + \frac{1}{2}$ (8) $\frac{1}{5} + \frac{3}{10}$

(9) $\frac{7}{10} + \frac{1}{5}$ (10) $\frac{3}{10} + \frac{1}{2}$ (11) $\frac{2}{5} - \frac{1}{5}$ (12) $\frac{3}{5} - \frac{2}{5}$

(13) $1 - \frac{4}{5}$ (14) $\frac{4}{5} - \frac{2}{5}$ (15) $\frac{3}{10} - \frac{1}{10}$ (16) $\frac{1}{2} - \frac{2}{5}$

(17) $1 - \frac{2}{5}$ (18) $\frac{7}{10} - \frac{3}{5}$ (19) $\frac{9}{10} - \frac{4}{5}$ (20) $\frac{7}{10} - \frac{2}{5}$

PROBLEMS

105

(1) Find one-third of 21.

(2) Mary spent $\frac{1}{3}$ of her money in a shop. What fraction of her money has she left?

(3) Six girls share a prize. What fraction do they each get?

(4) How many fifths are equal to eight-tenths?

(5) Bob and Charles each get quarter of an orange. They put their quarters together. What part of the orange does this make?

(6) Tom lost $\frac{1}{5}$ of his stamps. What fraction has he left?

(7) A cake was divided into 8 equal slices. I ate 3 slices. What fraction did I eat?

(8) How many tenths must I add to $\frac{3}{10}$ to get a whole one?

(9) Find $\square$:
(a) $\frac{1}{4} = \frac{\square}{8}$ (b) $\frac{1}{2} = \frac{\square}{8}$ (c) $\frac{3}{4} = \frac{\square}{8}$.

(10) How many new pence are there in $\frac{1}{2}$ of 50p?

(11) John had an apple. He halved it and then he halved it again. What is the name of each part?

(12) What is the value of $\frac{1}{2}$p + $\frac{1}{2}$p?

(13) Find $\square$: Four-fifths are $\square$ times one-fifth.

(14) I saw $\frac{3}{4}$ of a film. What fraction did I miss?

(15) A rectangle has 24 boxes. How many boxes are there in $\frac{1}{6}$ of the rectangle? $\frac{5}{6}$?

(16) Jane had a bar of chocolate. She gave me $\frac{1}{4}$ of it and then $\frac{1}{8}$ of it. What fraction did I get?

(17) Draw a rectangle and shade $\frac{7}{10}$ of it.

(18) What part of one-half is one-sixth?

(19) How many minutes are there in $\frac{1}{4}$ of an hour? $\frac{3}{4}$ hour?

(20) Mother gave me $\frac{5}{8}$ of the sweets and told me to give baby $\frac{1}{8}$ of the sweets. What fraction have I left?

Decimal Notation

Make your own abacus.

You need five rods and a number of rings to fit the rods.

Each rod can be fitted into a cardboard box.

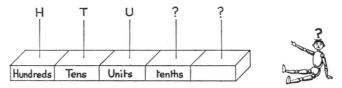

(1) How many units are there in one ten?

(2) How many tens are there in one hundred?

(3) What fraction of ten is one unit?

(4) Copy and complete this sentence:

One ten = $\frac{1}{10}$ of _____

(5) If one unit is divided into ten equal parts what is the name of one part?

(6) Write one tenth as a fraction.

(7) How many rings on the 'Tens' rod are equal to one ring on the 'Hundreds' rod?

(8) Copy and complete

_____ T equals 10 U

(9) Which rod would take a ring equal to $\frac{1}{10}$ unit?

(10) How many rings would you place on the 'tenths' rod to show $\frac{3}{10}$.

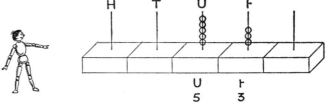

Abe's abacus shows

Abe reads five units and three-tenths.

Note: The place value immediately to the right of units (U) is named tenths (t).

The number on the abacus is written $5\frac{3}{10}$.

> $5\frac{3}{10}$ can be written 5·3
>
> 5·3 is read five 'point' three

The 'point' separates units place from 'tenths' place and is called a decimal point.

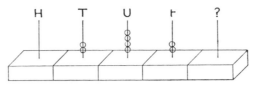

Abe sees 2 Tens, 3 Units and 2 tenths on his abacus.
He writes 23·2.
He reads "Twenty three 'point' two".

Read and write the numbers shown on each abacus:

107

(1)   (2)

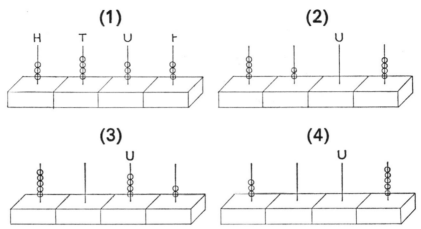

(3)   (4)

Show each of these numbers on an abacus sketch:

HTU t

(5) 635·4   (6) 707·2   (7) 206·8
(8) 100·4   (9) 823·3   (10) 542·1

(11) How many rings on the 't' rod equal one ring on U rod?

(12) How many rings on the 'h' rod equal one ring on 't' rod?

(13) How many rings on the 'h' rod equal one ring on U rod?

A ring on the 't' rod has a value of $\frac{1}{10}$ of a unit.
$\frac{1}{10}$ of a unit is written 0·1
A ring on the 'h' rod has a value of $\frac{1}{100}$ of a unit.
$\frac{1}{100}$ of a unit is written 0·01

Copy and complete:

(14) 0·46 is shown on an abacus, by □ rings on the 'tenths' rod and □ rings on the 'hundredths' rod.

(15) Six rings on the 'tenths' rod and five rings on the 'hundredths' rod is written ⌐ 0·   ⌐

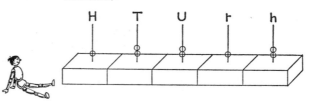

Abe reads "One Hundred and Twenty-two 'point' one two".

Abe writes 122·12.

The point is called a Decimal Point and separates Units place from tenths place.

Fractions and Decimals

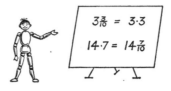

$$3\tfrac{3}{10} = 3 \cdot 3$$
$$14 \cdot 7 = 14\tfrac{7}{10}$$

Write as decimals:

108

(1) $\tfrac{3}{10}$ (2) $\tfrac{6}{10}$ (3) $\tfrac{7}{10}$ (4) $\tfrac{5}{10}$ (5) $\tfrac{9}{10}$

(6) $\tfrac{1}{10}$ (7) $1\tfrac{3}{10}$ (8) $3\tfrac{7}{10}$ (9) $4\tfrac{5}{10}$ (10) $2\tfrac{9}{10}$

Write in decimal form:

(11) Two 'point' six (12) Four 'point' seven
(13) Ten 'point' four (14) Twenty four 'point' eight
(15) Seventy 'point' seven (16) Sixty-two 'point' five
(17) $28\tfrac{7}{10}$ (18) $123\tfrac{4}{10}$
(19) $320\tfrac{3}{10}$ (20) $205\tfrac{7}{10}$

Write these decimals using fractions:

(21) 2·3 (22) 0·3 (23) 0·9 (24) 6·7
(25) 0·8 (26) 5·9 (27) 12·7 (28) 72·2
(29) 14·6 (30) 21·4 (31) 80·9 (32) 141·7

$\tfrac{1}{100} = 0 \cdot 01$
$\tfrac{7}{100} = 0 \cdot 07$
$\tfrac{39}{100} = 0 \cdot 39$

$$\tfrac{9}{100} = 0 \cdot 09$$
$$0 \cdot 03 = \tfrac{3}{100}$$

Write in decimal form:

109

(1) $\tfrac{7}{100}$ (2) $\tfrac{9}{100}$ (3) $\tfrac{3}{100}$ (4) $\tfrac{5}{100}$ (5) $\tfrac{6}{100}$

(6) $\tfrac{43}{100}$ (7) $\tfrac{27}{100}$ (8) $\tfrac{49}{100}$ (9) $\tfrac{78}{100}$ (10) $\tfrac{57}{100}$

(11) $2\tfrac{37}{100}$ (12) $4\tfrac{39}{100}$ (13) $19\tfrac{21}{100}$ (14) $23\tfrac{33}{100}$

Write as decimals:

(15) Two 'point' four (16) Three 'point' one four
(17) Nine 'point' one seven (18) Twelve 'point' three six
(19) Sixty four 'point' four five (20) Twenty one 'point' seven

Decimal Addition and Subtraction

24 + 32

$$\begin{array}{r} T\ U \\ 2\ 4 \\ +\ 3\ 2 \\ \hline 5\ 6 \end{array}$$

t for 'tenths'

$$\begin{array}{r} U\cdot t \\ 2\cdot 4 \\ 3\cdot 2 \\ \hline 5\cdot 6 \end{array}$$

2·4 + 3·2

KEEP the decimal points underneath each other

4·63 + 6·75 + 2·21

h for 'hundredths'

$$\begin{array}{r} U\cdot t\ h \\ 4\cdot 6\ 3 \\ 6\cdot 7\ 5 \\ 2\cdot 2\ 1 \\ \hline 13\cdot 5\ 9 \end{array}$$

Addition:

(1) $\begin{array}{r} 1\ 4 \\ 1\ 2 \\ 1\ 3 \\ \hline \end{array}$

(2) $\begin{array}{r} 1\cdot 4 \\ 1\cdot 2 \\ 1\cdot 3 \\ \hline \end{array}$

(3) $\begin{array}{r} 5 \\ 3 \\ 1 \\ \hline \end{array}$

(4) $\begin{array}{r} 0\cdot 5 \\ 0\cdot 3 \\ 0\cdot 1 \\ \hline \end{array}$

(5) $\begin{array}{r} 2\cdot 2 \\ 3\cdot 1 \\ 4\cdot 2 \\ \hline \end{array}$

(6) $\begin{array}{r} 7\cdot 4 \\ 2\cdot 2 \\ 1\cdot 2 \\ \hline \end{array}$

(7) $\begin{array}{r} 2\cdot 3 \\ 1\cdot 2 \\ 6\cdot 2 \\ \hline \end{array}$

(8) $\begin{array}{r} 14\cdot 21 \\ 21\cdot 32 \\ 10\cdot 24 \\ \hline \end{array}$

(9) $\begin{array}{r} 3\ 5 \\ 3\ 7 \\ 2\ 3 \\ \hline \end{array}$

(10) $\begin{array}{r} 3\cdot 5 \\ 3\cdot 7 \\ 2\cdot 3 \\ \hline \end{array}$

(11) $\begin{array}{r} 12\cdot 3 \\ 27\cdot 7 \\ 32\cdot 8 \\ \hline \end{array}$

(12) $\begin{array}{r} 4\cdot 62 \\ 2\cdot 13 \\ 5\cdot 26 \\ \hline \end{array}$

(13) 0·6 + 0·3

(14) 0·8 + 0·1

(15) 0·4 + 0·2 + 0·1

(16) 1·2 + 1·3 + 1·4

(17) 2·3 + 1·5 + 3·1

(18) $\begin{array}{r} 3\cdot 14 \\ 2\cdot 23 \\ 1\cdot 52 \\ \hline \end{array}$

(19) $\begin{array}{r} 4\cdot 26 \\ 7\cdot 35 \\ 2\cdot 41 \\ \hline \end{array}$

(20) $\begin{array}{r} 6\cdot 35 \\ 4\cdot 26 \\ 2\cdot 34 \\ \hline \end{array}$

(21) 3·25 + 4·65

(22) 6·14 + 3·25 + 2·62

Find □ :

(23) 0·6 + 0·3 = □

(24) 1·2 + 0·7 = □

(25) 1·3 + 1·5 = □

(26) 3·45 + 3·12 = □

110

Addition:

(27) 23·61	(28) 0·13	(29) 72·34
34·23	7·00	21·34
16·05	3·64	40·03

(30) 61·62 + 14 + 3·23

PROBLEMS

III

(1) Add two 'point' five, three 'point' six-four and four 'point' six.
(2) Find the answer to 79·06 + 48 + 81·98.
(3) Add 39·81, 24·06 and 85·38.
(4) Find the answer to 74·08 + 25·65 + 41·64.
(5) Add 143·2, 232·4 and 226·2.
(6) Add $2\frac{7}{10}$ to 5·8.
(7) Use decimal notation to add $4\frac{3}{10}$ and $5\frac{4}{10}$.
(8) Find the answer to $2\frac{1}{10} + 4\frac{5}{10} + 3\frac{3}{10}$.
(9) Add 2·36, 1·27 and 16·32.
(10) Find the answer to 28·16 + 54·03 + 63·28.
(11) Add $14\frac{17}{100}$, $2\frac{33}{100}$ and $12\frac{7}{100}$.
(12) Add 31·24, 41.23, 26·82 and 14·61.

ABE'S EXERCISE

Addition:

112

(1) £1·20
£1·31
£   ·

(2) 3·25 marks
2·32 marks
·   marks

(3) 5·15 dollars
3·43 dollars
dollars

(4) 2·35 litres
4·62 litres
litres

(5) 6·43 jumblies
2·34 jumblies
jumblies

(6) 14·62 umpties
23·81 umpties
umpties

(7) 15·66 francs
28·06 francs
francs

(8) £25·62
£31·35
£

(9) 21·36 metres
9·54 metres
metres

(10) 16·830 kilogrammes
14·210 kilogrammes
kilogrammes

SUBTRACTION OF DECIMALS

86 − 23   8·6 − 2·3

```
    2 3      8 6          2·3      8·6
   +6 3     −2 3         +6·3     −2·3
    8 6      6 3          8·6      6·3
```

Subtraction:

(1) 0·4 − 0·2   (2) 0·8 − 0·5   (3) 0·14 − 0·11

(4)   8·5   (5)   9·6   (6)   6·7   (7)   3·7
     −3·4   −6·4   −0·5   −2·0

(8)   1·3   (9)   6·2   (10) 10·8
     −0·8   −4·3   −3·9

Copy and fill in the missing decimal number in each:

(11) 8·9 − □   = 2·3   (12) □   − 4·3 = 5·3

(13) □ − 17·5 = 7·1   (14) 49·8 − 24·6 = □

(15) 33·7 − □   = 23·3   (16) 79·76 − 38·24 = □

Subtract:

(17) 12·34   (18) 125·34   (19) 84·50
      5·55   96·25   79·35

(20) 23·9   (21) 168·65
     12·72   93·36

(22) Find the answer to 17·83 + 3·18 − 2·5

(23) What number added to 14·52 gives 20·68?

(24) 20·68 − 14·52 = □

ABE'S MIXTURES

Find the answers to:

(1) (4·6 + 2·3) − 5·6   (2) 7·3 + 2·7 − 5

(3) 29·45 + 6·21 − 9·45   (4) 17·8 + 0·2 − 7·5

(5) Add 8·62 to 3·18 and then subtract 7·23.

(6) Add $15\frac{3}{10}$ to $17\frac{7}{10}$.

(7) A roll of cloth is 20·25 metres long and 6·15m of cloth is sold. What length of cloth is left?

(8) A man has £6·25 and spends £4·15. How much has he left?

(9) What must I add to 18·76 to get 23·24?

(10) Find the difference between fourteen 'point' seven six and seventeen 'point' five four.

British Coins

50p   10p   5p   2p   1p $\frac{1}{2}$p

£1 = 100p

50p is read as fifty new pence or one 'fifty'
10p is read as ten new pence or one 'ten'
5p is read as five new pence or one 'five'
2p is read as two new pence
1p is read as one new penny
$\frac{1}{2}$p is read as half of one new penny

£2·48$\frac{1}{2}$ is read as two pounds forty eight and a half new pence

Two $\frac{1}{2}$p	= 1p
Five 1p	= 1 'five'
Two 'fives'	= 1 'ten'
Five 'tens'	= 1 'fifty'
Two 'fifties'	= £1
Ten 'tens'	= £1
Twenty 'fives'	= £1
100p	= £1

Find the answers to:

115

(1) $3\frac{1}{2}$p + 2   (2) $7\frac{1}{2}$p + $2\frac{1}{2}$ (3) $11\frac{1}{2}$p + $23\frac{1}{2}$
(4)   £1·24$\frac{1}{2}$ + £2·25$\frac{1}{2}$ (5)   £6·31$\frac{1}{2}$ + £8·27$\frac{1}{2}$
(6)   How many 'fives' are equal to one 'fifty'?
(7)   One 'ten' is equal to how many 1p coins?
(8)   How many $\frac{1}{2}$p coins would you give for one 2p coin?
(9)   How many 2p coins would you give for one 'ten'?
(10)   A man has a 'fifty', four 'tens' and two 'fives'. How much money has he?
(11) $20\frac{1}{2}$p + 5p + $6\frac{1}{2}$p   (12) $42\frac{1}{2}$p + $21\frac{1}{2}$p + $31\frac{1}{2}$p
(13)   £2·64 + £5·13$\frac{1}{2}$ (14)   £12·66 + £37·82$\frac{1}{2}$

Money Addition and Subtraction

You have been adding decimals.
Now, we will add money in the *same* way.
Keep the points underneath each other like
the Snowman's buttons.

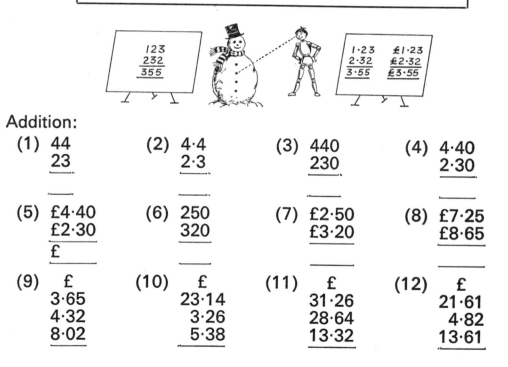

116 Addition:

(1) 44
       23
       ___

(2) 4·4
      2·3
      ___

(3) 440
       230
       ___

(4) 4·40
      2·30
      ___

(5) £4·40
      £2·30
      £____

(6) 250
       320
       ___

(7) £2·50
      £3·20
      ____

(8) £7·25
      £8·65
      ____

(9) £
      3·65
      4·32
      8·02
      ____

(10) £
       23·14
        3·26
        5·38
       _____

(11) £
       31·26
       28·64
       13·32
       _____

(12) £
       21·61
        4·82
       13·61
       _____

Find the answers to:

(13) £2·24 + £3·35

(14) £11·36 + £12·24

(15) Add 'four pounds twenty four' to 'six pounds sixteen'.

(16) Find the answer to £17·35 + £13·25 + £18·32.

(17) Add 16·24 francs, 29·61 francs and 13·23 francs.

(18) Add 41·26 kroner, 21·82 kroner and 6·23 kroner.

MONEY PROBLEMS

Copy and fill in the missing sum of money in each box:

117

    (1) £46·24 + £2·42 = ☐

    (2) £71·35 + £16·28 = ☐

    (3) £14·62 + £6·23 = ☐

    (4) Find the answer to £18·65 + £19·35.

    (5) Find the answer to £71·23 + £82·65.

> £1 = 100p

    (6) Add 24p, 38p and £1.

    (7) What is the total sum of three pounds sixteen and forty new pence?

    (8) A boy has one 'fifty' coin and two 'tens'. How many new pence has he?

    (9) A conductor collects one 10p fare, one 5p fare and one 2p fare. How much did he collect?

 (10) A carpet cost £72·16 and the charge for 'fitting' the carpet was £6·82. What was the total cost?

 (11) A grocer's bill is made up of three items, whose prices are £2·34, £1·25 and 75 new pence. What is the total bill?

 (12) A weekly household bill is:
Grocer £4·25, Butcher £2·32 and Milk £1·42. What is the total bill?

 (13) Add £3·62, 50p and 80p.

 (14) A transistor costs £11·62 and the batteries cost 40p. How much did the transistor, with batteries, cost?

SUBTRACTION OF MONEY

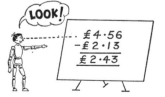

Subtraction

118

(1) 524	(2) £5·24	(3) £6·98	(4) £7·24
−212	−£2·12	−£4·72	−£0·20
———	£————	———	———

(5)	£	(6)	£	(7)	£14·76 − £12·34
	21·36		33·25		
	−15·89		−26·38		

(8) £18·94 − £12·71   (9) £23·24 − £11·12   (10) £77·81 − £55·63

PROBLEMS

(1) A man has £5 and spends £3·50. How much has he left?

(2) A lady has three £1 notes and a 'fifty' coin in her purse. She spends £2·25. How much has she left?

(3) A tea room bill is seventy-six new pence. How much change is given from £1?

(4) A man spends £8·65 out of £10. How much money has he left?

(5) Find the answer to £14·64 + £3·38 − £8·35.

(6) Add £3·75 to £5·25 and take away £6·50.

(7) Abe has two pounds sixty and he is given one pound forty-five. How much does he now have?

(8) The grocer's bill is £2·25 and the milk bill £1·63. How much money is needed to pay both bills?

(9) At the end of term the teacher gave Abe one pound and told him to buy 30 packets of sweets at 3p each. How much change should Abe get?

(10) From £5·74 subtract £2·48 and add £6·74 to the answer.

Length

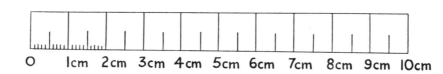

O   1cm 2cm 3cm 4cm 5cm 6cm 7cm 8cm 9cm 10cm

The ruler shows **10** centimetres.

REMEMBER

> 100 cm = 1m

> 1 metre twenty-five centimetres is written 1·25m
> 2·36m is read as two 'point' three six metres or
> two metres thirty-six centimetres
> m is read as metres; cm is read as centimetres

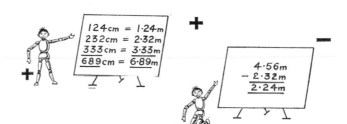

Find the answers to:

120

(1) 1·32m
   +2·23m
   —————

(2) 4·35m
   +3·24m
   —————

(3) 6·71m
   +2·28m
   —————

(4) 4·62m
   +5·36m
   —————

(5) 23·75m
   +7·25m
   —————

(6) 24·36m
   +21·85m
   —————

(7) 4·35m
   +0·23m
   +1·31m
   —————

(8) 12·30m + 25·62m + 3·04m

(9) 14·04m + 16·16m + 23·25m

(10) 2·23m
    −1·23m
    —————

(11) 7·59m
    −4·35m
    —————

(12) 8·99m
    −2·28m
    —————

(13) 9·98m	(14) 36·21m	(15) 123·20m
−5·36m	−21·85m	−85·60m

PROBLEMS

121
(1) 23m + 13·25m + 28·64m
(2) 4·35m − 1·52m
(3) 100m − 37·5m
(4) 28m + 25cm + 4·05m
(5) (73·75m + 26·25m) − 37·5m
(6) 14·62m + 5·38m − 6·81m
(7) 71·35m − 56·89m + 28·65m
(8) 46cm + 25cm − 0·31m

ABE ASKS QUESTIONS

122

(1) How many cm are there in one metre?
(2) How many centimetres are there in 5m?
(3) How many cm are there in 1·50 metres?
(4) Copy and complete: 2·5m = □ cm.
(5) A 'fifty' coin is equal to □ new pence. Find □.
(6) How many new pence are there in £1?
(7) How many new pence are there in one 'ten' coin?
(8) Write 250p in £.
(9) How many 'fifties' are equal to £4?
(10) Which is the longer length 200cm or 2 m?

ABE'S PUZZLERS

123
(1) A roll of silk measures six metres and 2m are sold. What is now the length of the roll?
(2) A boy ran in a 60m race and received a 5m start. How far did he actually run?
(3) A man buys 3·54m of wood for 100p and another 2·46m fo 45p. How much wood did he buy and how much did it cost
(4) A boy is 1·34m tall. What is his height in cm?
(5) What is the length of your classroom in metres?
(6) Take six 'point' seven eight metres from eleven 'point' two fiv metres.
(7) A boy has 5·25m of wood and he makes two shelves eac 1·25m long. How much wood has he left after making th shelves?
(8) A strip of wood is 9m long. How many 3m lengths can b cut from it?
(9) A 'running' track has two 80m 'bends' and two 120 metr 'straights'. What is the distance round the track?
(10) How tall is teacher?

Weight

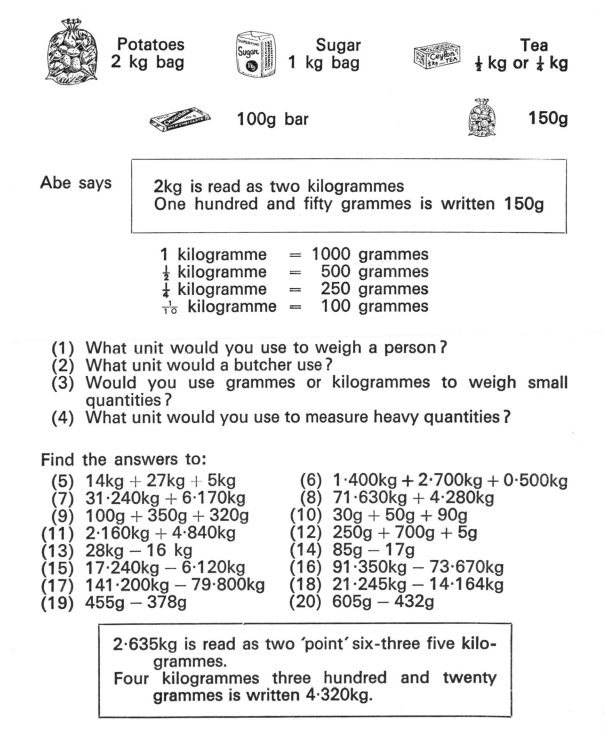

Potatoes
2 kg bag

Sugar
1 kg bag

Tea
½ kg or ¼ kg

100g bar

150g

Abe says

| 2kg is read as two kilogrammes |
| One hundred and fifty grammes is written 150g |

1 kilogramme	=	1000 grammes
½ kilogramme	=	500 grammes
¼ kilogramme	=	250 grammes
$\frac{1}{10}$ kilogramme	=	100 grammes

124

(1) What unit would you use to weigh a person?
(2) What unit would a butcher use?
(3) Would you use grammes or kilogrammes to weigh small quantities?
(4) What unit would you use to measure heavy quantities?

Find the answers to:

(5) 14kg + 27kg + 5kg
(6) 1·400kg + 2·700kg + 0·500kg
(7) 31·240kg + 6·170kg
(8) 71·630kg + 4·280kg
(9) 100g + 350g + 320g
(10) 30g + 50g + 90g
(11) 2·160kg + 4·840kg
(12) 250g + 700g + 5g
(13) 28kg − 16 kg
(14) 85g − 17g
(15) 17·240kg − 6·120kg
(16) 91·350kg − 73·670kg
(17) 141·200kg − 79·800kg
(18) 21·245kg − 14·164kg
(19) 455g − 378g
(20) 605g − 432g

| 2·635kg is read as two 'point' six-three five kilo- |
| grammes. |
| Four kilogrammes three hundred and twenty |
| grammes is written 4·320kg. |

Picture Graphs

125 **POCKET MONEY**

chocolate   chewing gum   sweets   toffees

This picture shows how a boy spends some of his pocket money.

(1) Count the pennies in each pile and write down:
    (a) the amount spent on chocolate;
    (b) the amount spent on toffees.

(2) Which is the highest pile of pennies?

(3) Which is the lowest pile of pennies?

(4) How much more is spent on toffees than on chewing gum?

126 Each pupil in the class should put his book on one of four bundles according to his birthday month. For example:

BIRTHDAYS

SPRING SUMMER AUTUMN WINTER
(Mar-May) (Jun-Aug) (Sep-Nov) (Dec-Feb)

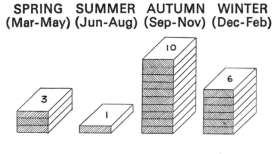

Look at the piles of books you now have *for your own class* and answer these questions:

(1) What is the most popular time of the year for birthdays in your class?

(2) Which time of year has least birthdays?

(3) Is there any season in which there are no birthdays?

Pupils wearing the smallest size of shoe in the class should put their books in one bundle. Pupils wearing the next size place their books in the next bundle, and so on until every pupil has his book in a bundle according to his size in shoes.

Here is a picture of the books for one class:

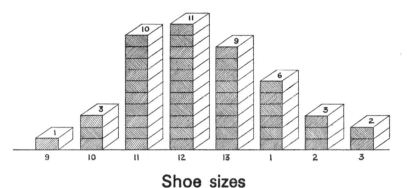

Shoe sizes

Note: In mathematics such a picture is called a BLOCK GRAPH.

Each pile of books (or blocks) is spaced evenly along the line across the page.

To make it easier to draw a block graph only the *outline* of each pile of blocks (or books) may be drawn. Coloured strips of paper cut to size could be used for these outlines:

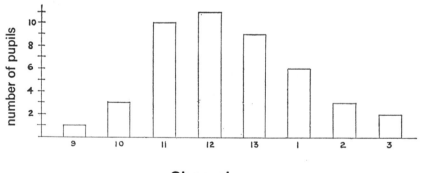

Shoe sizes

Answer these questions from the graph above:

127
(1) What size of shoe occurs most often in this class?

(2) How many pupils wear shoes smaller than size 12?

(3) How many pupils wear shoes larger than size 12?

Use the 'piles of books' for the pupils in your own class and draw an *outline* block graph. Answer the questions above from your own graph.

Here is a block graph showing how high some pupils **can reach** when standing up on tip-toe:

HEIGHT PUPILS CAN REACH

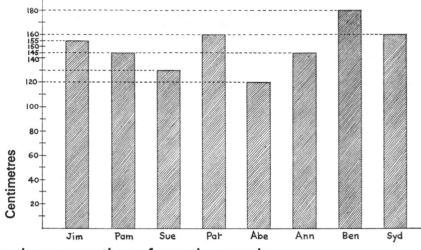

Answer these questions from the graph:

128

(1) Which pupil can reach highest?
(2) How high can this pupil reach?
(3) How high can Jim reach?
(4) Which two pupils can reach up to 145 centimetres?
(5) How high can Pat and Syd reach?
(6) Which pupils can reach higher than Pam?

Copy and complete this table for your class:

Number in Family	0	1	2	3	4	5	6	7 or more
Number of Pupils								

Draw an outline block graph to show these numbers.
Answer these questions from your graph:

129

(1) In your class what is the most common size of family?
(2) How many pupils have more than 5 in the family?

Multiplication by 'Tens'

Copy and complete:

130

(1) $12 \times 2 \times 10 = 24 \times 10 = 240$ (2) $12 \times 20 = 240$
(3) $23 \times 2 \times 10 = 46 \times 10 =$ (4) $23 \times 20 =$
(5) $32 \times 3 \times 10 = \quad \times 10 =$ (6) $32 \times 30 =$
(7) $31 \times 4 \times 10 = \quad \times 10 = 1240$ (8) $31 \times 40 =$
(9) $30 \times 5 \times 10 = \quad \times 10 =$ (10) $30 \times 50 =$
(11) $43 \times 5 \times 10 = 215 \times 10 =$ (12) $43 \times 50 =$
(13) $55 \times 6 \times 10 = \quad \times 10 =$ (14) $55 \times 60 =$
(15) $63 \times 7 \times 10 = \quad \times 10 =$ (16) $63 \times 70 =$
(17) $72 \times 8 \times 10 = \quad \times 10 =$ (18) $72 \times 80 =$
(19) $85 \times 9 \times 10 = \quad \times 10 =$ (20) $85 \times 90 =$

60×86 Multiply by 6 tens
$$\begin{array}{r} 86 \\ \times 60 \\ \hline 5160 \end{array}$$
     86×60

Find the products:

131

(1) 21×50 (2) 32×40 (3) 40×20 (4) 53×30
(5) 62×40 (6) 74×20 (7) 83×30 (8) 91×70
(9) 22×60 (10) 27×20 (11) 35×40 (12) 36×50
(13) 44×80 (14) 53×60 (15) 67×70 (16) 69×70
(17) 75×50 (18) 87×80 (19) 89×60 (20) 96×90

Multiplication:

132

(1) 20×36 (2) 30×25 (3) 30×44 (4) 40×56
(5) 50×57 (6) 60×43 (7) 70×81 (8) 70×47
(9) 80×59 (10) 90×14

PROBLEMS

133

(1) There are 20 chocolates in each box. How many are in 14 boxes?

(2) There are 60 minutes in 1 hour. How many minutes are in 24 hours?

(3) A shopkeeper hands in 32 'Fifties' to the bank. How many new pence should he get for them?

(4) Find the product of 65 and 80.

(5) A motorist buys 40 litres of petrol each week. How many litres of petrol does he buy in a year of 52 weeks?

Multiplication by Two Digits

Copy and complete:

134

(1)
$$22 \times 20 = 440$$
$$+22 \times 3 = 66$$
$$\overline{22 \times 23 = 506}$$

(2)
$$31 \times 30 = 930$$
$$+31 \times 1 =$$
$$\overline{31 \times 31 = 961}$$

(3)
$$42 \times 40 =$$
$$+42 \times 2 =$$
$$\overline{42 \times 42 = 1764}$$

(4)
$$51 \times 40 =$$
$$+51 \times 5 =$$
$$\overline{51 \times 45 = 2295}$$

(5)
$$54 \times 50 =$$
$$+54 \times 4 =$$
$$\overline{54 \times 54 =}$$

(6)
$$66 \times 60 =$$
$$+66 \times 3 =$$
$$\overline{66 \times 63 =}$$

(7)
$$68 \times 60 =$$
$$+68 \times 7 =$$
$$\overline{68 \times 67 =}$$

(8)
$$75 \times 70 =$$
$$+75 \times 5 =$$
$$\overline{75 \times 75 =}$$

(9)
$$80 \times 80 =$$
$$+80 \times 6 =$$
$$\overline{80 \times 86 =}$$

(10)
$$97 \times 80 =$$
$$+97 \times 8 =$$
$$\overline{97 \times 88 =}$$

Tens first

$$\begin{array}{r} 98 \\ \times 96 \\ \hline 8820 \\ 588 \\ \hline 9408 \end{array}$$

98×96

Find answers, like Abe, multiplying the 'tens' first:

135

(1) 26×25	(2) 41×52	(3) 28×32	(4) 44×64
(5) 84×71	(6) 45×16	(7) 54×92	(8) 48×41
(9) 41×82	(10) 69×53	(11) 22×86	(12) 46×36
(13) 39×14	(14) 55×48	(15) 23×74	(16) 81×63
(17) 57×27	(18) 59×45	(19) 52×67	(20) 26×94

Units first

$$\begin{array}{r} 76 \\ \times 54 \\ \hline 304 \\ 3800 \\ \hline 4104 \end{array}$$

76×54

Find answers, like Abe, multiplying the 'units' first:

136

(1) 62×73	(2) 34×46	(3) 22×61	(4) 34×76
(5) 64×85	(6) 18×96	(7) 35×18	(8) 16×37

(9) 77×56	(10) 84×78	(11) 14×65	(12) 62×46
(13) 44×39	(14) 43×58	(15) 32×29	(16) 75×24
(17) 84×59	(18) 93×88	(19) 73×34	(20) 99×99

Find the products:

137

(1) 62×21	(2) 38×51	(3) 52×42	(4) 81×72
(5) 22×15	(6) 47×17	(7) 54×19	(8) 89×23
(9) 98×35	(10) 45×38	(11) 63×43	(12) 74×49
(13) 57×46	(14) 66×57	(15) 72×68	(16) 36×79
(17) 55×83	(18) 64×87	(19) 72×95	(20) 84×97

PROBLEMS $\times$ and $\div$

138

(1) There are 24 hours in a day and 7 days in a week. How many hours are there in a week?

(2) Teacher showed us some colour slides. She put them on at the rate of 4 to a minute. The show lasted 16 minutes. How many slides were shown?

(3) A case holds 180 eggs. How many eggs are in 9 cases?

(4) The sweet shop sells toffees in packets of 8. How many packets can be filled from 1072 toffees?

(5) 224 men played rugby. They were arranged in teams of 7. How many teams were there?

(6) The butcher makes up sausages in packets of 9. How many sausages will he need to make up 106 packets?

(7) A cafe can seat 236 people. Each table has 4 chairs around it. How many tables are in the cafe?

(8) Abe's problem

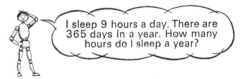

I sleep 9 hours a day. There are 365 days in a year. How many hours do I sleep a year?

(9) There were 19 groups dancing at a ball. There were 8 people in each group. How many people were dancing?

(10) Mother bought 1000 grammes of meat for 8 of us to share. How many grammes should we each get?

(11) A bus can carry 76 passengers. How many passengers can 14 buses carry?

(12) Each classroom in a school has 42 pupils' desks. How many pupils' desks are in the school, which has 22 classrooms?

Metric Units

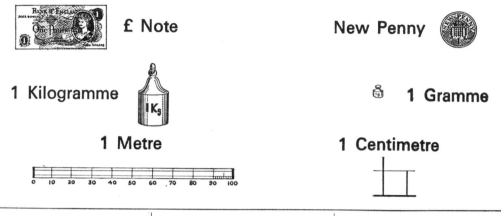

£ Note   **New Penny**

1 Kilogramme   **1 Gramme**

1 Metre   **1 Centimetre**

Symbol	Name (Read)	Numeral (Write)
£	Pound	£1·00
p	New Penny	1p or £0·01
kg	kilogramme	1 kg
g	gramme	1g
m	metre	1m
cm	centimetre	1cm or 0·01m

> 100p = £1
> 1000g = 1kg
> 100cm = 1m

Read:

  (1) 4·02m   (2) £6·25   (3) 4·240kg   (4) 25·270g
  (5) 61·23m   (6) £15·39

Write in numerals:

  (7) Fourteen pounds fifty four.
  (8) Sixteen 'point' seven six five kilogrammes.
  (9) Fifty-one 'point' seven two five metres.
 (10) Ten pounds and seventy-six new pence.

Decimal Multiplication

$$3 \times 4{\cdot}2$$

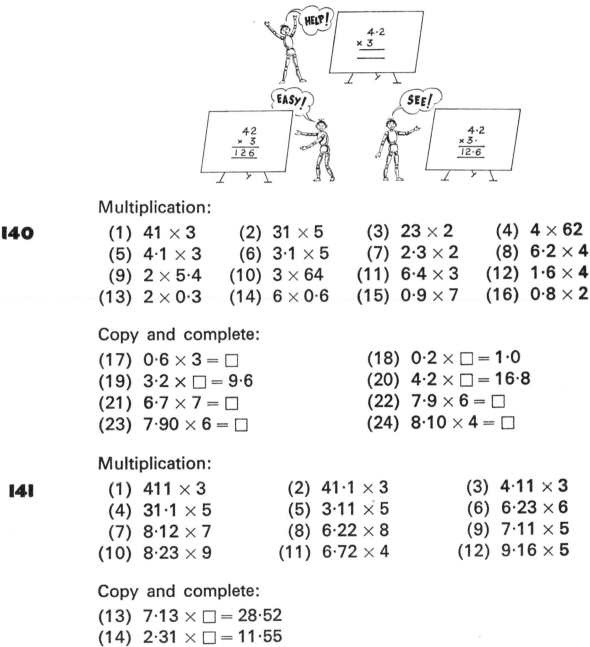

Multiplication:

140

(1) 41×3 (2) 31×5 (3) 23×2 (4) 4×62

(5) $4{\cdot}1 \times 3$ (6) $3{\cdot}1 \times 5$ (7) $2{\cdot}3 \times 2$ (8) $6{\cdot}2 \times 4$

(9) $2 \times 5{\cdot}4$ (10) 3×64 (11) $6{\cdot}4 \times 3$ (12) $1{\cdot}6 \times 4$

(13) $2 \times 0{\cdot}3$ (14) $6 \times 0{\cdot}6$ (15) $0{\cdot}9 \times 7$ (16) $0{\cdot}8 \times 2$

Copy and complete:

(17) $0{\cdot}6 \times 3 = \square$ (18) $0{\cdot}2 \times \square = 1{\cdot}0$

(19) $3{\cdot}2 \times \square = 9{\cdot}6$ (20) $4{\cdot}2 \times \square = 16{\cdot}8$

(21) $6{\cdot}7 \times 7 = \square$ (22) $7{\cdot}9 \times 6 = \square$

(23) $7{\cdot}90 \times 6 = \square$ (24) $8{\cdot}10 \times 4 = \square$

Multiplication:

141

(1) 411×3 (2) $41{\cdot}1 \times 3$ (3) $4{\cdot}11 \times 3$

(4) $31{\cdot}1 \times 5$ (5) $3{\cdot}11 \times 5$ (6) $6{\cdot}23 \times 6$

(7) $8{\cdot}12 \times 7$ (8) $6{\cdot}22 \times 8$ (9) $7{\cdot}11 \times 5$

(10) $8{\cdot}23 \times 9$ (11) $6{\cdot}72 \times 4$ (12) $9{\cdot}16 \times 5$

Copy and complete:

(13) $7{\cdot}13 \times \square = 28{\cdot}52$

(14) $2{\cdot}31 \times \square = 11{\cdot}55$

(15) $7{\cdot}31 \times \square = 29{\cdot}24$

(16) $9{\cdot}45 \times \square = 47{\cdot}25$

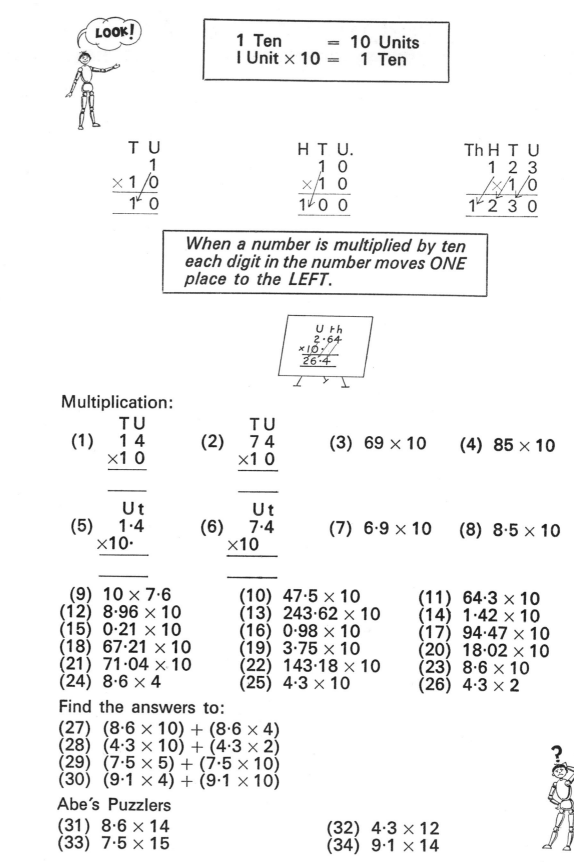

LOOK!

| 1 Ten = 10 Units |
| 1 Unit × 10 = 1 Ten |

```
  T U              H T U.            Th H T U
    1                  1 0              1 2 3
  ×1 0              × 1 0            × 1 0
  ─────            ───────          ─────────
  1 0              1 0 0            1 2 3 0
```

When a number is multiplied by ten each digit in the number moves ONE place to the LEFT.

```
  U th
  2·64
 ×10
 ─────
 26·4
```

142 Multiplication:

(1)
```
 T U
 1 4
×1 0
─────
```

(2)
```
 T U
 7 4
×1 0
─────
```

(3) 69 × 10

(4) 85 × 10

(5)
```
 U t
 1·4
×10·
─────
```

(6)
```
 U t
 7·4
×10
─────
```

(7) 6·9 × 10

(8) 8·5 × 10

(9) 10 × 7·6
(10) 47·5 × 10
(11) 64·3 × 10
(12) 8·96 × 10
(13) 243·62 × 10
(14) 1·42 × 10
(15) 0·21 × 10
(16) 0·98 × 10
(17) 94·47 × 10
(18) 67·21 × 10
(19) 3·75 × 10
(20) 18·02 × 10
(21) 71·04 × 10
(22) 143·18 × 10
(23) 8·6 × 10
(24) 8·6 × 4
(25) 4·3 × 10
(26) 4·3 × 2

Find the answers to:
(27) (8·6 × 10) + (8·6 × 4)
(28) (4·3 × 10) + (4·3 × 2)
(29) (7·5 × 5) + (7·5 × 10)
(30) (9·1 × 4) + (9·1 × 10)

Abe's Puzzlers
(31) 8·6 × 14
(33) 7·5 × 15
(32) 4·3 × 12
(34) 9·1 × 14

Progress Checks

PROGRESS CHECK 7

(1) Copy and write the correct sign for △:   54 △ 6 = 9
(2) Write the name of this number: 4260

(3) Write down the time on this clock:

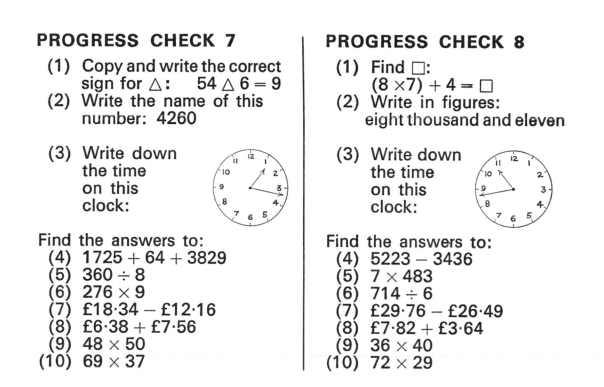

Find the answers to:
(4) 1725 + 64 + 3829
(5) 360 ÷ 8
(6) 276 × 9
(7) £18·34 − £12·16
(8) £6·38 + £7·56
(9) 48 × 50
(10) 69 × 37

PROGRESS CHECK 8

(1) Find □:
    (8 × 7) + 4 = □
(2) Write in figures: eight thousand and eleven

(3) Write down the time on this clock:

Find the answers to:
(4) 5223 − 3436
(5) 7 × 483
(6) 714 ÷ 6
(7) £29·76 − £26·49
(8) £7·82 + £3·64
(9) 36 × 40
(10) 72 × 29

PROGRESS CHECK 9

(1) Mother baked a cake and divided it into 6 equal slices. Jane ate 1 slice and Tom ate 2. What fraction of the cake was left?
(2) Dad's shirt costs £1·75. How much change should he get from £5?
(3) What is the total value of a 'Fifty', a 'Ten', a 'Five', a 'Two' and a 'One'?
(4) There are 7 days in a week. How many days are there in 52 weeks?
(5) Tom is 1·42m tall. Bill is 14cm taller than Tom. What is the height of Bill?
(6) There were 24 children at Sheila's party. ¼ of them were boys. How many girls were at the party?
(7) Peter left home for school at 25 past 8. He arrived in the playground at 7 minutes to 9. How long did it take him to go to school?
(8) Find the total cost of a skirt at £3·42 and a jumper at £1·78.

Extra Practice

Find these sums:

A

(1)	243	(2)	427	(3)	327	(4)	511	(5)	562
	301		361		314		106		173
	125		111		213		39		160

Find the answers to:

(6) 231+284+361   (7) 245+316+232   (8) 461+17+358
(9) 824+156+314   (10) 717+206+307   (11) 232+274+635
(12) 293+171+526   (13) 119+620+553   (14) 28+67+882
(15) 345+286+273   (16) 161+117+759   (17) 215+490+146
(18) 358+332+48   (19) 357+143+484   (20) 231+659+260

Find these differences:

B

(1)	684	(2)	745	(3)	896	(4)	764	(5)	884
	321		542		305		217		316

Find the answers to:

(6) 526 − 518   (7) 966 − 174   (8) 857 − 393
(9) 659 − 262   (10) 754 − 716   (11) 832 − 829
(12) 666 − 589   (13) 643 − 258   (14) 552 − 369
(15) 441 − 194   (16) 386 − 187   (17) 567 − 408
(18) 812 − 721   (19) 903 − 716   (20) 958 − 785

Find the answers to:

C

(1) 38×7 (2) 7×83 (3) 73×8 (4) 8×73 (5) 94×3
(6) 49×6 (7) 6×94 (8) 9×46 (9) 64×9 (10) 3×49
(11) 8×95 (12) 59×8 (13) 89×5 (14) 5×98 (15) 34×9
(16) 6×87 (17) 78×6 (18) 7×68 (19) 86×7 (20) 9×43

Find the answers to:

D

(1) 256×5 (2) 271×9 (3) 279×6 (4) 342×4
(5) 364×2 (6) 9×435 (7) 5×457 (8) 3×462
(9) 2×574 (10) 4×685 (11) 689×8 (12) 723×8
(13) 754×8 (14) 798×7 (15) 758×9 (16) 9×826
(17) 7×847 (18) 934×8 (19) 5×919 (20) 938×6

Find the answers to:

E
(1) $138 \div 2$ (2) $176 \div 4$ (3) $574 \div 7$ (4) $738 \div 9$
(5) $936 \div 2$ (6) $936 \div 6$ (7) $912 \div 6$ (8) $920 \div 8$
(9) $573 \div 3$ (10) $584 \div 4$ (11) $735 \div 5$ (12) $912 \div 8$
(13) $987 \div 7$ (14) $804 \div 3$ (15) $630 \div 5$ (16) $980 \div 7$
(17) $490 \div 2$ (18) $906 \div 3$ (19) $917 \div 7$ (20) $776 \div 8$

Find the quotients and remainders to:

F
(1) $128 \div 6$ (2) $146 \div 3$ (3) $164 \div 5$ (4) $182 \div 8$
(5) $189 \div 2$ (6) $255 \div 4$ (7) $265 \div 6$ (8) $274 \div 7$
(9) $342 \div 5$ (10) $418 \div 9$ (11) $436 \div 3$ (12) $504 \div 5$
(13) $520 \div 7$ (14) $575 \div 8$ (15) $621 \div 2$ (16) $636 \div 9$
(17) $738 \div 4$ (18) $840 \div 9$ (19) $896 \div 6$ (20) $913 \div 4$

Find the answers to:

G
(1) $(4 \times 8) + 5$ (2) $(5 \times 7) + 2$ (3) $(3 \times 9) + 6$
(4) $(6 \times 6) + 8$ (5) $(2 \times 10) - 3$ (6) $(7 \times 7) - 6$
(7) $(8 \times 6) - 4$ (8) $(8 \times 9) - 2$ (9) $(7 \times 9) + 7$
(10) $(8 \times 10) + 12$ (11) $(9 \times 5) - 14$ (12) $(9 \times 9) - 11$
(13) $(8 \times 8) + 7$ (14) $(9 \times 4) - 8$ (15) $(8 \times 0) + 7$
(16) $(9 \times 7) - 23$ (17) $50 - (8 \times 5)$ (18) $40 - (7 \times 4)$
(19) $(9 \times 10) + (2 \times 5)$ (20) $(8 \times 3) + (4 \times 5)$

Find the sums:

H

(1)
```
  1416
  2130
  3213
```

(2)
```
  432·3
  213·1
  104·2
```

(3)
```
  2142
  1448
  1397
```

(4)
```
  12·72
  24·13
  41·32
```

Find the answers to:

(5) $1425 + 2166 + 452$ (6) $2453 + 3663 + 1125$
(7) $3262 + 2721 + 1664$ (8) $3588 + 1196 + 2574$
(9) $45·89 + 14·33 + 27·73$ (10) $375 + 5159 + 1606$
(11) $234·8 + 138·2 + 328·1$ (12) $162 + 1787 + 3337$
(13) $14·21 + 1·55 + 25·43$ (14) $3697 + 2345 + 2169$
(15) $35·37 + 25·67 + 10·15$ (16) $2284 + 3574 + 1404$
(17) £6·36 + £9·28 + £8·64   (18) £376 + £896 + £723
(19) £9·45 + £30·77 + £24·27   (20) £4252 + £1268 + £176

Find the answers to:

I

(1)
```
  2846
 -1242
```

(2)
```
  29·52
 -17·10
```

(3)
```
  3753
 -1312
```

(4)
```
  386·4
 -241·6
```

(5)
```
  3458
 -3129
```

(6)
```
  47·13
 -13·06
```

(7)
```
  4652
 -2271
```

(8)
```
  47·38
 -34·46
```

(9) 5555 −2280	(10) 516·6 −383·6	(11) 5279 −1472	(12) 638·4 −255·1
£ (13) 6823 −3236	£ (14) 62·46 −44·17	£ (15) 7437 −1562	£ (16) 75·91 −46·22
£ (17) 8314 −3456	£ (18) 82·03 −26·25	£ (19) 9431 −5678	£ (20) 90·06 −71·58

Find the answers to:

J

(1) 18×30	(2) 23×40	(3) 25×20	(4) 26×20
(5) 28×50	(6) 29×50	(7) 36×30	(8) 36×40
(9) 43×60	(10) 52×70	(11) 57×80	(12) 61×40
(13) 61×50	(14) 68×70	(15) 76×80	(16) 78×80
(17) 83×90	(18) 92×90	(19) 97×50	(20) 97×60

Find the answers to:

K

(1) 28×21	(2) 32×24	(3) 36×29	(4) 45×32
(5) 51×35	(6) 53×38	(7) 58×43	(8) 61×46
(9) 64×47	(10) 67×52	(11) 69×55	(12) 72×58
(13) 74×61	(14) 78×64	(15) 83×66	(16) 86×73
(17) 89×77	(18) 92×84	(19) 95×89	(20) 97×96

B.1.